621.4
McD   McDonald, Lucile
        Windmills: an old-new energy
        source

This book may be kept
FOURTEEN DAYS
A fine will be charged for each
day the book is kept overtime.

1/9			
11-5			
12-2			

MEDIALOG
Alexandria, Ky 41001

84
83
82

Windmills

WINDMILLS

An Old-New
Energy Source

by **Lucile McDonald**

Line drawings by Helen Hawkes Battey

Elsevier/Nelson Books
New York

Library of Congress Cataloging in Publication Data

McDonald, Lucile Saunders.
Windmills: an old-new energy source.
Bibliography: p.
Includes index.
SUMMARY: Describes windmills through the
ages and in various countries.
1. Windmills—History—Juvenile literature.
[1. Windmills—History] I. Title.
TJ823.M25   621.4'5   80–22680
ISBN 0–525–66708–3

Published in the United States by
Elsevier-Dutton Publishing Co., Inc.,
2 Park Avenue, New York, N.Y. 10016.
Published simultaneously in Don Mills,
Ontario, by Nelson/Canada.

Printed in the U.S.A.   First Edition
10 9 8 7 6 5 4 3 2 1

To my son, Dick,
who inspired this book
and who has had a part in it
from the beginning

Contents

chapter 1

How Mills Began

The world is now rediscovering one of man's oldest sources of energy—the wind. When electricity and steam took over the task of furnishing power for the mechanical aspects of everyday life, windmills fell into disuse. Today however, fossil fuels—coal and oil—are in shorter supply. Once removed from the ground, they are not replaceable; they are gone forever. Therefore, mankind is looking for substitutes. And while some industrialists look to nuclear and geothermal means to solve our energy problem, others are thinking once again of harnessing the wind.

The story of how the wind's force was first applied to human needs goes back at least a thousand years and into ancient lands. Early people thought there was something uncanny about putting the wind to work and were slow to learn it was possible. They were accustomed to doing their tasks the hard way, by hand.

Yet in time there were hundreds of thousands of windmills scattered around the globe, doing such chores for man as pumping water, grinding flour, and sawing wood. Many of them added beauty to the landscape, but as the centuries went by and windmills were modernized, they lost their romantic look. Instead of picturesque sails and graceful towers, they became metal fans spinning at the top of skinny iron pyramids. Today's windmills are stark propellers mounted high on reinforced concrete poles or steel towers. And if wind machines once were picturesque additions to the landscape, today people worry about whether they will clutter the environment.

Wind power is clean, free, and inexhaustible, but it alone cannot meet the energy demands of the modern world. But there are vast regions of the earth where it can supplement other available power sources, and that is why efforts are being made to bring windmills back into common use.

Men discovered how to sail boats with the wind before they found other ways to put it to work. Primitive farmers carried water from their streams to their fields by hand. They worked all day in the hot sun, irrigating the small plots of ground they had plowed with oxen and a pointed stick. There were few inventions to help the farmer. He grew what grain he could, harvested it with his primitive tools, then took it to his stone hut or skin tent for his wife and daughters to grind it into meal.

Before grain was cultivated, the earliest people ate berries and herbs and nuts. Finding that acorns could be crushed into flour, they tried giving other hard kernels the same treatment. Having learned that wheat and other grains were nourishing, they puzzled over means to get at the tiny food particles, but the only method they could devise was pounding the kernels between rocks.

Then someone thought of spreading the wheat on a smooth stone, shaped something like a saddle, and rubbing a long rock back and forth over it, the way some Mexican women still prepare their corn for tortillas. The crushed grain was boiled in a porridge and patted into cakes.

Statuettes have been found in Egyptian tombs showing women kneeling to grind grain on such stones.

This was a slow way to provide bread for the earth's steadily increasing population. The grinding process was improved when an early-day inventor fitted a smooth rock into a hollow stone, made a hole for grain to be poured into the contrivance, and added a stick for a handle to move the inner stone back and forth. This grinder was known as a quern, and it was the type of mill used about the time of Christ.

Commercial milling began with the Romans and the next step up was their hourglass-shaped mill, large stones with bars extending from the sides so that men could walk around and around, turning them. This kind of labor was usually done by slaves.

Grain had to be ground the world over, and it was considered a most unpleasant task. There is a folk story in Ireland of a third-century king, Cormac Ulfada, who did not love his wife and yearned instead for the Princess Ciarnute, daughter of the king of the Picts in Britain. When the queen heard of her husband's unfaithfulness, she had Ciarnute seized and sentenced to labor at the mills until she ground nine measures of corn each day. Soon the girl became ill and begged Cormac to save her. The king feared his wife's power, so he deceived her and sent to Scotland for a workman who, it was said, could make an engine to grind grain. That is how Irish minstrels came to sing of their island country's first mechanical mill, built to rescue Princess Ciarnute.

As the world's requirements for flour grew, slave mills could not grind enough grain, so mules and horses were put to work, which could make the stones revolve faster. The stones were increased in size, and draft animals turned them night and day. Next millers conceived the idea of placing the stones upstairs in a building and connecting them with a toothed cogwheel and a central shaft attached to the pole. The animals turned the stones by walking around the pole on the lower floor.

Though the quern was the most common type of mill, other kinds of mills had come into existence. Probably man's first use of stones connected with gears was in the water wheel. Someone in Asia Minor or in Greece discovered that the current of a stream could push millstones faster than a horse could do it. The first water wheels stood upright like a turnstile, and the grinding stones were directly above the paddles. As the blades moved, a pole fastened in the center turned the top stone. The invention of the cogged wheel permitted the paddle to turn horizontally in the stream, as a modern water wheel does today.

Mountain streams were not always available for powering mills, and the inhabitants of low, sandy countries needed something else. These places often had an abundance of wind, and in time someone found a way to harness its power. Sails had been used to power boats ever since the Egyptians, four thousand years before Christ, hoisted a single square of linen on masts and sailed their craft down the Red Sea.

Other nations had employed the wind in various ways. Tibetans had learned to send their prayers spinning off to heaven on crude little prayer wheels that whirled in the wind. An inventive Greek in Alexandria tinkered with an unusual instrument, an Aeolian harp, played by the wind passing over its strings. It was time for someone to find a means to use the wind to irrigate the thirsty soil and grind grain for food.

chapter 2

The First Wind Machines

The earliest ideas regarding a practical wind machine came out of the Middle East.

In the first half of the seventh century A.D. an Arabian caliph named Omar governed the city of Medina. He had heard rumors from Iran, then called Persia, of a curious invention that could grind grain with the wind. But he could learn nothing more about this mysterious machine until he was told of a slave who was reported to have seen it. A certain Abu Lulu was a carpenter who had been taken prisoner by the Greeks in his youth and was later captured by Muslims at the battle of Nehavend. He had been put up for sale in the marketplace of Medina, and the man who bought him took the lion's share of his earnings at carpentry.

According to Arab histories, one day when Omar was passing in the street, Abu Lulu broke away from his master and, casting himself at the caliph's feet, cried, "O Commander of the Faithful, right my wrong. Verily hath my master taxed me heavily."

"At how much?" asked the caliph.

"At two dirhems a day."

"And what is thy trade?"

"Carpenter and worker in iron," the slave replied.

"It is not much for a clever artificer like thee," said Omar. "Are you not the man I have heard about who could design for me a mill driven by the wind?"

The slave admitted that this was true, but his eyes glowed angrily.

"Come, then," urged the caliph. "I would see such a marvel. Make for me this wind machine."

Abu Lulu rose defiantly. He had hoped that by accosting Omar he might be granted his liberty. "If spared," he muttered grimly, "I will make a mill for thee, a mill whose fame shall reach from the Far East unto the Far West."

The man's tone sounded more like a threat than an agreement, but Omar was a tolerant ruler, and Abu Lulu could do him an important service. He permitted the slave to go his way, not knowing that the carpenter's thoughts were fixed on revenge because he had been a captive almost all of his life.

Next day, at the hour of morning prayer, Caliph Omar joined the worshipers in the great mosque. He was not aware that Abu Lulu was in the front row, squatting on a prayer mat. Omar took his place and knelt before the holy niche, his face toward Muhammad's birthplace, his back toward the congregation.

"*Allah il allah*," commenced the chant, as foreheads touched the floor.

At that instant Abu Lulu sprang to his feet, drew a knife from his girdle, and plunged it deep into Omar's back. Six times he stabbed the caliph before he struck the murderous blade into his own heart.

This happened in the year 644 A.D. Abu Lulu carried the secret of the wind machine to his grave. For three hundred years not another word was heard of the mysterious invention.

In the tenth century there lived a famous Arab historian and geographer called Masudi. Born in Baghdad, he grew to manhood with a great desire to acquire a store of knowledge. He learned to read and write quickly, and packed his head with all sorts of information picked up from the astrologers, geographers, and scientists of the court. Yet Masudi thirsted for more. To him the earth was a source of hidden mines filled with curious and precious facts that could not be found in the pages of books.

When he was a young man, Masudi followed the caravan trails to lands unknown to his people, making notes on what he saw. Eventually these jottings filled thirty books, an encyclopedia of useful information, which he titled *Meadows of Gold and Mines of Precious Stones.* In it he told of sea serpents, pearl fishing, the habits of the rhinoceros, the intrigues of kings, and so many other subjects that scholars have called him the Herodotus of the Arabs.

In the course of his travels Masudi visited northeast Iran, where he crossed a desert in the province of Seistan. It was a tiring journey, for on the vast, dry plain a strong gale blew in the same direction for months at a time. The wind flung harsh particles of sand in the traveler's face and parched his skin. Masudi was relieved when he reached the city of Zaranj, lying amid gardens on the shore of a salty inland sea. In these gardens he saw something marvelous he had never seen before—engines worked by wind that ground grain and helped with other work. The people said the mills were very old, and that an Egyptian warrior had brought them to their country in the time of Moses.

Masudi wrote down all he could learn about them and called the machines windboards. That was a very suitable name, since they did not have arms whirling around in the

A Seistan windmill. The Arab historian Masudi called these windmills wind-boards when he discovered them in Iran in the tenth century. They were low on the ground and had a vertical shaft like a revolving door. The wheel was inside the tower, which had an opening for the wind to pass through.

Photo courtesy of Professor Grant M. Farr

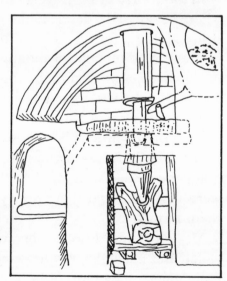

Line drawing reprinted from Traditional Crafts of Persia, *by Hans Wulff, by permission of The MIT Press, Cambridge, Massachusetts.*

air, but were upright and low on the ground. They had a vertical shaft like a revolving door, and a wall of bricks was placed around half the structure, leaving an open space for the wind to blow through from the other half. The windboards worked only as long as the wind blew from the right direction.

Today in the desert of Seistan, there is no city of Zaranj because in the fourteenth century the fierce Mongol conqueror Tamerlane destroyed the irrigation system and devastated the country in revenge for a wound received when he invaded it. The high winds that pushed the wind-boards moved the sand dunes along until they buried the gardens and what was left of the town. Inhabitants moved away to more fertile places, and only desolation remained. Most Iranians have forgotten that this remote corner of their country may have been the birthplace of the windmill.

After Masudi's mention of this mechanical wonder, there was another long gap in the world's knowledge of the wind machine. Word went forth slowly over the caravan trails that sails could be fitted to a cogged shaft, harnessing the wind in the same manner that water wheels had conquered the swift currents of streams. Eagerly men listened to these tales, and each, according to his understanding, tried to apply the information to local conditions. In the Crimea, the Russians placed their sails low on the ground and turned them horizontally. In Palestine (Israel) they were erected high on fortress walls, thus ensuring the inhabitants a supply of flour in times of siege. There were many other variations.

During the twelfth and thirteenth centuries, warriors were traveling from Europe to the Holy Land to fight the Crusades. Along such highways as existed people were curious to know what returning knights and their followers had seen. Was it true there existed such marvels as lemons

A tiny windmill perched on the battlements of a Crusader fortress.

and sugar? Could the Arabs really look at themselves in glass mirrors? What was damask like, and lilac dyes and cloth made of camel's hair? And did strange machines pushed by the wind really grind grain?

Yes, said the Crusaders, there were such wonders. A voyager might see winged monsters standing on breezy hilltops or along arms of the land reaching toward the sea, wherever the wind blew strongest.

Anyone journeying to the islands off the coast of Asia Minor could see those mills. There were scores of stone towers with sails found on islands in the Aegean Sea —Tenedos, Mykonos, Stampalia (Astypalaia), and Rhodes. In 1403, the Spanish ambassador to the court of Tamerlane passed by Rhodes and counted fourteen windmills near the harbor. When the Spanish diplomat saw these towers grinding flour and crushing olives for oil, they looked to him as though they had been working for several centuries. Each had been given the name of a patron saint, such as the Windmill of the Virgin or the Windmill of St. Catherine.

In 1309, after the Crusades, the Knights Hospitallers of St. John of Jerusalem, who had been driven out of the Holy Land, settled on this pleasant, fertile island, which grew

Farmers on the island of Rhodes preferred windmills with eight cloth sails. Some are still in use in the Greek islands.

almost everything its people needed. Every community owned a windmill, a stone tower with eight triangular cloth sails like those of a sailing craft, a concept that was natural to maritime people. Where the Colossus of Rhodes had stood in ancient times (before an earthquake tumbled the gigantic statue into the sea), two breakwaters reached out into the blue Aegean, and alongside them were so many towers with spinning arms that the place was known as the Harbor of the Windmills. They were no doubt the ones the Spanish ambassador later reported.

All was peaceful on Rhodes until the Turkish sultan Suleiman the Magnificent declared war upon the knights and came with a fleet to besiege the island. The defenders were expecting him and posted lookouts to give warning when the enemy appeared. Meanwhile the crops of barley and wheat ripening in the fields were speedily harvested. Horses and burros laden with grain plodded day and night over narrow trails to the nearest windmills, where men worked inside at the grinding stones, sacking flour to be carried to underground storerooms inside the city walls. Peasants brought their cattle and poultry with them, and cellars were well stocked in preparation for a siege. By the time the first of the Turkish fleet heaved into view, the inhabitants were safely ensconced inside their stone battlements.

Unceasingly the enemy battered the gates with rams and hurled stone balls at the round towers, but the islanders were safe behind their staunch walls, fortified with their best defense weapon, plenty of food. The siege lasted five months, during which munitions and other supplies finally were exhausted. Nevertheless, the Grand Master had no intention of surrendering until one day the sultan himself, unescorted except for an interpreter, appeared at the principal gate and offered to let the knights leave the island honorably with all the possessions they wished to take. The weary defenders accepted and set sail for Italy on January 1, 1523.

The old mills of the harbor withstood the fighting that went on around them, and some are still there, inhabited by local families. These structures are different from those of Europe and are scarcely higher than the nearby palm trees. No taller mills were needed to catch the winds that blow on the Rhodian shore.

chapter 3

Windmills of the Middle Ages

When wind machines reached Europe in the twelfth century, they were mistakenly called "Turkish mills." This probably was because they came out of the east, as did the nomadic Turks who just then were moving from Asiatic regions down into the valleys of Asia Minor, where they were soon to build the Ottoman Empire.

Northern European windmills were not at all like those the Crusaders may have seen, and it is believed they were an independent invention. Germans claimed that the scythe and some other useful agricultural implements originated in central Europe, and it was believed that the first windmills were built on mountaintops in Bohemia. It is true that a Bohemian made the first crude drawing of the inside workings of a windmill that has ever been found, but there is no evidence that his countrymen had anything to do with inventing the windmill itself. France is known to have had a windmill in 1105 at Arles, another in Normandy in 1180, and there is a documentary reference to one in an English village in Yorkshire in 1185. Six years later the story of the Abbot Samson and another English mill was told in *The Chronicle of Jocelien of Brackelonde,* a parchment volume the monks copied by hand. It spoke of the days when the wind belonged only to rich landowners and the peasants had no more rights than slaves. Noblemen and bishops owned everything, sometimes even the village bread oven and winepress, so that the poor could not prepare food and drink without paying their overlord for it. Barons controlled the horse mills and water mills, and the peasants gave part of each sack of grain to pay for grinding their flour. This portion, taken out for the master, was called the miller's "golden thumb."

According to the story in the parchment manuscript, a kind old dean at the monastery of Bury St. Edmonds heard of this new invention, the wind machine that could drive millstones as fast as a hundred men could turn them by hand. This strange and wonderful creation did not have to be fed, nor did it ever become ill or complain. Deciding to have such a machine for his parish, the dean called in a carpenter and the two pooled their knowledge, guessing at the way to build what was to become known as a peg-and-post mill.

It consisted of a tiny house set on a central support and light enough so that a man could pick up a pole and turn the whole building by hand if the wind changed. Such a windmill often was built around an oak tree, which served as a shaft as well as a sturdy anchor if a storm blew up. When a tree stump was not used, the millhouse was set on a wooden tripod, held to the earth with pegs. A post mill required an ingenious carpenter because the house not only had to balance and pivot, but also be able to withstand strong winds and the vibration that went with grinding. If it could not be turned when the wind shifted, the sails would not whirl and work would stop. Some early windmills were put on boats for this reason, because it was easy to raise the anchor and turn the whole mill on the water.

A tripod braced on the ground was the support for some early mills.

At that point, no one had yet devised a way to set the sails into the wind without twisting the whole building around. Therefore, the first European mills that appear in pictures had a house not more than twice the size of a horse, with room inside it for only one man to work. The earliest known drawing of one appeared in the Windmill Psaltery copied at Canterbury in 1270.

This was the kind of modest structure that the dean of Bury St. Edmonds had built. It was barely finished when the Abbot Samson heard about it and immediately concluded that this crazy invention would take money out of his pocket. All the peasants would flock to Bury St. Edmonds with their grain, if for no other reason than to see the strange machine at work.

Seething with anger, the abbot ordered workmen sent to destroy the contraption and store the wood for kindling. Dean Herbert heard what was about to happen and hastened to plead with the abbot to withdraw his command.

"You might as well cut off both my feet as build such a mill," the abbot declared. "The wind belongs to the abbey, and you cannot use it. Until that wicked machine is destroyed I will eat no bread."

The old dean gave up and hurried home to have his own carpenter take the mill apart. He knew that in the eyes of

Here a sturdy tree trunk made the best kind of anchor for a primitive post mill.

the law, Samson was right; the wind was not the dean's to use—it belonged to the rich and powerful abbot.

The same thing happened in many other countries. From the time windmills first appeared, manorial prerogatives included the right to permit or refuse building them, thus compelling tenants to have their grain ground at the lord's mill. Another right enabled the lord to prohibit buildings or trees to be placed where they might interfere with a free sweep of the wind.

Much luckier than the English dean were the monks of the abbey at Windsheim in Holland, who planned to erect a windmill at Zwoll. The powerful lord of that part of the Netherlands claimed that the wind in the entire district was his. But the monks did not give up easily. They appealed to the Bishop of Utrecht and he supported them, saying that the wind belonged to him and they could go ahead with their mill.

After several centuries, the windmills of northern Europe evolved from the peg-and-post type into simple structures that had their "feet in the earth," as the millers said, so that

This most curious old English mill wore a long skirt and had its "foot in the ground," as the millers said.

they would not blow over easily. At first the tripod was enclosed in a small house, on top of which the mill was perched.

These structures often were architectural wonders, for they were built with as few metal fastenings as possible and yet were exceedingly strong. The windmills of northern Europe generally had four sails and a post extending toward the ground, so that it could be turned into the wind. Years later, the post was given a wheel on the lower end to make it easier to push around.

The life of a miller was not an easy one, for he had to take advantage of the hours when the wind blew, no matter what time it was. This has always been a problem with wind power; it is unpredictable and irregular. The miller might be awakened from sound sleep and work for the next two days and nights without stopping. There was one rule, however: no matter how hard the wind blew, Sunday was a day of rest.

chapter 4

English Windmills

Windmills, usually being the tallest, most conspicuous buildings on the landscape in medieval times, played an important part in history, especially in Great Britain, where there were once ten thousand of them. A story is told about one, the Mill of the Hide, sometimes called King Harry's Mill.

During a civil war King Henry III was defeated by the army of the barons at the battle of Lewes in 1264. His younger brother, Richard, who called himself King of the Romans, was unable to flee after the fight. He took refuge in the windmill, propping heavy millstones against its door. His pursuers readily tracked him to the little mill and stood outside shouting taunts.

"Come out, you paltry miller," they called. "You, forsooth to turn a wretched millmaster! You who defied us and would have no meaner title than King of the Romans! Come out!"

Richard could not endure their ridicule, so he opened the door and gave himself up.

Many kings found windmills a safe lookout point from

This drawing of an English windmill is taken from a piece of sculptured brass made about 1350 and put in St. Margaret's Church at King's Lynn.

which to watch a fight. At the battle of Crécy in 1346, King Edward III climbed into the tower of a stone windmill with walls seven feet thick. He couldn't have found a better observation post.

Charles I in 1642 watched the first fight between the Roundheads and Cavaliers from a post mill at Edgehill. Three years later, from another mill, he saw his troops defeated at Naseby.

Charles II had a different kind of experience connected with a windmill. He had just lost the battle of Worcester in 1651, and his enemies were trying to find him. It was a dark, cold night when he fled in the company of a friend. The nearest shelter was the Evelith windmill, and the two men headed for it stealthily. Before they realized they had been observed, the miller threw open the door and demanded, "Who goes there?"

The king's companion shouted, "Neighbors going home."

The miller was suspicious, for the voice was not familiar.

"If you be neighbors," he replied, "stand, or I will knock you down."

The fugitives feared the miller intended to delay them while he summoned help, so they fled down the muddy lane,

An open trestle post mill of the earliest type found in England. It was one like this that sheltered King Charles I.

slipping and stumbling in the dark, until they found a place where they could jump a hedge. Once on the other side, they lay breathless in a ditch.

The miller came after them and searched the bushes in vain, shouting, "Rogues! Rogues!" At length he gave up and took his lantern back to the mill. Then the two men continued on their way.

Long afterward the king learned that the miller was his friend and would have offered shelter. The enemy troops so thoroughly suspected the miller of concealing the king that they turned the Evelith windmill topsy-turvy searching for him. All they found for their trouble was an odd little verse well known to millers:

> *By God's fair air*
> *I grind ye grain,*
> *Make good prayer*
> *When bread ye gain.*

Charles was not yet through with his windmill adventures. On his way to the coast friends lent him a horse that usually carried wheat and flour to the windmill at Whiteladies. When his pursuers were close on his trail, the miller hastily helped him hide in a hollow oak tree and went back to work.

The leader of the searching troops came up and offered the miller a thousand pounds if he would tell him where Charles was hiding.

"He must have passed by another road," the dusty master of Whiteladies suggested.

When the pursuers were out of sight, the miller helped the king out of the oak tree and he finished his journey safely.

Sherwood Forest, where legends tell us Robin Hood and his merry men once lived, had windmills. One of them was kept by John Cockle, who was both forester and miller. One night Cockle heard a gun being fired in the woods and, after some search, found a richly clad, portly gentleman alone in the timber. The stranger said he was a wayfarer who had become separated from his friends, but Cockle did not believe him. He was convinced the man was a poacher.

Nevertheless, Cockle took the fellow to the mill and permitted him to stay all night in the flour-coated old tower. In the morning he showed him the right road out of the forest.

Later that same day Cockle was out again after what he supposed was a party of poachers. He insisted on bringing them to the mill, where they did a great deal of protesting. Couldn't he see that they were the king's courtiers and they were out looking for Henry VIII, who was lost in the forest? Cockle told them of his portly guest, whereupon they declared he had bedded the king for the night.

Cockle might have doubted the story if King Henry had not sent for him, thanked him officially for the night's lodging, made him a knight, and given him a pension for life.

England had a region of marshy and pond-covered land known as the fens, which was inhabited by primitive people who derived most of their living from fishing and hunting water fowl. In 1634 the government hired a Dutch engineer, who brought skilled workers to drain the region so it would be suitable for agriculture, since Netherlanders were experienced in handling drainage problems. The Hollanders were to be paid with a certain share of the reclaimed land.

A host of pumping windmills sprang up as they progressed, but the sight of these angered the fen dwellers. Seeing their way of life threatened, they carried on open warfare, breaking down embankments and wrecking sluices. Much as they tried to discourage the Dutch engineer, he kept on with his project until 1655 and was fairly successful, although floods and storms caused setbacks. It was the nineteenth century before the drainage in that part of England was completed and the pumping windmills that accomplished the task disappeared.

chapter 5

Don Quixote's Windmills

While England and northern Europe got along as best they could with post mills, a quite different type of mill appeared in Spain with the coming of the Moors.

Like those of Rhodes, the Moorish windmills were built of stone with conical roofs. Their four lengthy arms barely cleared the earth. The Arab tribesmen had so long ago harnessed the wind that they knew how to utilize it for other purposes than just grinding grain and lifting water. They established Europe's first paper mill at Játiva and powered it with sails.

The stone towers with whirling wings had become fairly common all over Spain when Miguel de Cervantes wrote *Don Quixote* (1605–1615), which contains the world's best-known windmill story.

Cervantes' hero, the clownish Don Quixote, spent so much time reading about bold knights and fair ladies that he decided to practice chivalry himself. For a horse he had only a raw-boned nag called Rosinante, which means "once a plow horse." Needing a lady love, he chose a plump neighborhood farm girl, whom he renamed Dulcinea. Next

One artist's concept of Don Quixote's mill. The real ones in this part of Spain had more arms and thicker, shorter towers.

he appointed a squire, because every knight must have a squire. He chose a dumpy farmhand named Sancho Panza and promised to make him governor of the first kingdom they conquered.

The don put on his heavy armor, mounted the plow horse and set forth, Sancho following on a long-eared donkey. They had not gone far on the road when Quixote pointed to a row of thirty or forty giants strung out in a line.

"I shall encounter and slay them, and with their spoils we shall enrich ourselves, for it is fair war and a good service to sweep such evil fellows off the face of the earth," he cried.

"Those aren't giants; they're windmills," Sancho told him.

"It is clear," the don shouted, "that thou art not experienced in the business of adventure. They are giants, and if thou art afraid, betake thyself to prayer, while I engage them in fierce and unequal combat."

With that, he rode to attack the nearest mill.

A stiff breeze was blowing and the first swinging sail tore the lance from the don's scrawny hands. The second sail tossed him to the ground. As Sancho helped his bruised and angry master to his feet, Quixote exclaimed, "These giants have been changed into windmills."

His very first adventure spoiled, he rode off, vowing revenge.

Along a barren mountain ridge near Campo de Criptana in central Spain stands what is left of a row of ancient windmills. One of them may be the remains of the sturdy giant that defeated Don Quixote and enriched our language with the phrase "tilting at windmills."

chapter 6

Dutch
Windmills

Many people think that the Dutch invented windmills because they are so much a part of the Netherlands countryside. In actuality, there were no mills in Holland until about 1200, when a few were used for grinding. Another two hundred years went by before they were adapted to their most important service—making the country fit for human habitation. Windmills are credited with creating Holland, helping to work the land and providing industries to sustain it.

Until about 1000 A.D. Holland consisted mostly of marshes with small, sluggish streams. The whole expanse was separated from the sea by a belt of sand dunes, and from time to time the sea would rush in and flood large areas. Some years floods swallowed entire towns.

This watery country, even today, lies several feet below sea level. Historians have recorded the Dutch people's incessant fight to reclaim their land with dikes, but little has been said about the part played by the windmill in this struggle.

Toward the end of the fourteenth century, there lived in

the village of Alkmaar in northern Holland a prosperous blacksmith and farmer called Florent Alkmade. He was weary of the perpetual struggle to save his crops from the ravages of wind and water. The dikes had already been built, but there was nothing to prevent rivers and creeks from overflowing the land.

It so happened that during a visit to Paris Alkmade met a man who had been to Bohemia. This new acquaintance told him about a new device in use there, a wheel set up in the air that produced power to pump water from a well.

Alkmade thought a great deal about the Frenchman's tale. At that time, windmills in Europe were used mainly for grinding grain. Other uses to which the Arabs had put them were still unknown in northern lands.

Alkmade had respect for the wind because he had sailed the breezy lagoons of Holland and seen it push good-sized boats. The more he thought about the wind pumping water from wells in Bohemia, the more convinced he became that it could pump flood water from his own land. Upon his return home, Alkmade secretly began to fashion a small wheel with sails. For a long time he could not get them to set at right angles to the wind, but at last he believed he had mastered the problem. But when he showed his model to his neighbors they laughed and said he was a fool to waste his time making toys.

Not in the least discouraged, Alkmade built a larger wheel with four sails and went on to figure out how to make the mechanism lift water. He completed his invention in 1408, and soon pumping windmills were in use in every part of Holland. For a long time their sails turned only in the northwest breeze, the wind that most often prevailed in the country, but other engineers worked on the problem and eliminated this handicap.

A large percentage of the mills that sprang up all over

The smallest and simplest mills of Holland were those scattered throughout the polder fields to pump water into the canals for drainage purposes.

Holland were simply for the purpose of draining the land and had no other work to do. Each was equipped with a revolving scoop that rested on the waterlogged ground. As a wind-driven shaft turned it, the scoop picked up water and emptied it into a ditch at a higher level. Sometimes scoop mills were necessary in series, each windmill lifting the water a little higher, until at last it was discharged into a ring canal and carried out to sea. This sounds like a slow process for even a large mill, but with many hundreds of mills whirling day and night, more and more land was reclaimed. Some of the drainage mills were equipped with an Archimedes screw instead of scoops. This turned continuously, bringing up water as the spiral trough revolved.

Holland began to flourish, and its ships brought in an abundance of foreign wares, causing new industries to spring up. Since the wind was the only natural source of energy available in the country, and there was an abundance of it, windmills soon were everywhere, engaged in a variety of tasks. By 1600 they hulled rice, pressed oil from seeds, sawed planks, ground chalk, and made rags into

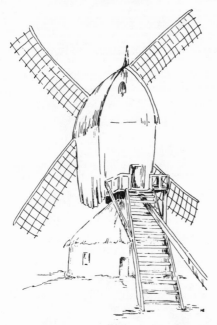

Later the post mill was closed in at the base and so became a "turret" mill. This one from the north of Holland is heavily thatched with brownish-green reeds and has a saddle roof.

paper, for which the invention of printing from movable type had created an enormous demand. The mills also pulverized snuff, cocoa, limestone, mustard, dyes, pepper, tanning bark, and all sorts of grain, and contributed to the processing of cloth. Before long, almost everybody depended upon windmills for a living. This probably explains why Holland had more mills of a greater variety than could be found anywhere else in so small an area. For centuries, water mills and windmills were the only complex machines in existence.

The Dutch did everything possible to improve their mills. It took a long time and much experimentation to find out what kind of mill was best adapted to any given part of the country. The Dutch were the first to put a wheel on the end of the beam so it would be easier to move it. They closed in the base of post mills, supplying a room for storage or work. Families even lived in the space, though the quarters were so cramped that the beds had to be built into the walls.

Formerly sawmills like this were common in Holland. Some of them could be turned completely around on the great ring at the base.

Northern Holland has a few old mills that are thatched all the way to the top with a thick layer of brown-green reed. This was pretty to look at, but in a high wind the mill sometimes caught fire from friction and burned like a giant straw stick. Lightning also presented another threat to windmills until, in modern times, their owners were persuaded to install lightning rods.

In the south of Holland, windmills were generally tall towers of brick, varnished with linseed oil to keep out the dampness. They had cleverly revolving domes, representing a complete change in style. With this type of mill it was no longer necessary to turn the entire turret in order to set the sails into the wind and start the shaft revolving. Only the roof, with the blades attached to it, had to be pushed into position, and this was accomplished with a beam that moved easily on a circular track.

Post mills like this were pushed around into the wind by an automatic gear, which sometimes was attached to the steps.

Another invention in 1745 by an Englishman, Edmund Lee, made the wind itself do the pushing by putting a small wheel on the side opposite the sails. Coupled to a gear, it caused the mill to shift automatically with changes in the breeze. The fantail slowly pulled the cap around and was a valuable timesaver for the busy miller. It worked so well that the fan wheel was sometimes attached to the steps or beams of the old post mills and caused the whole top of the structure to move. This was a lot different from the days when a man had to get out in stormy weather and push until the arms of his mill were in the wind.

There is a list of inventors who continued to make improvements, among them Andrew Meikle, a Scotsman, who in 1772 designed a sail to spill the wind during squalls. The sailcloth was replaced with a series of wooden shutters. Some mills had both kinds of sails. A more recent improvement was by A. Dekker of Leyden, who about 1930 began streamlining the sails, rounding and plating them with aluminum or zinc along the leading edge, which caused them to revolve in the faintest breeze. It was hoped at the time that this might make the use of windmills popular again, although power-driven pumps had already started to take their place.

Holland's golden era for windmills was around 1850, when there were nine thousand mills at work in this little country. They were of all sizes—from tiny ones beside the drainage ditches to great brick towers with a workshop and ample space for the miller's family on several floors. Zaandam alone had twelve hundred mills in its vicinity, used for all sorts of industries. Old men used to tell of the terrible noise the mills made on windy days, although for the people inside the mills the sounds consisted only of a low murmur—a combination of the rush of the circling sails, the clucking of wooden gears, the dry rustle of grain being poured in, and the whisper of the whirling stones.

It was different with the wind-powered sawmill brought to Zaandam by Cornelis Cornelison in 1592. At first it had no turning wheel, so he put the entire structure on a raft in the river. When the wind changed, up came the anchor and Cornelison moved the whole barge. Finally he put his saws in a trim, shapely tower on shore, which was promptly nicknamed the Flapper. There it stood for three hundred years, until 1891. A model of Holland's first wind-driven sawmill may be seen in a Dutch museum.

chapter 7

Windmills
Travel Far

$\mathbf{A}$s in the case of any practical invention, windmills were bound to travel to places other than European countries. The mills often took strange forms, but the purpose was the same—to put the wind to work for man.

Prisoners taken by the Mongol conqueror Genghis Khan are believed to have been familiar with the primitive mills in northeastern Iran. Through them, windmills were introduced in China to assist with irrigation. No towers were built; the sails were kept close to the ground and were often made of matting. They are described as vertical-axis mills, with the sails blowing around a central post connected with a scoop wheel to bring up the water.

A Chinese scholar, Wang Cheng, is given credit for having spread information about mills throughout the remote parts of China. Before the year 1620, when he was gathering notes for a book about inventions, he became acquainted with John Torrence, a Jesuit missionary, and questioned him about mechanical contrivances of the Western world. When Torrence told him how the wind had been harnessed, the Chinese scholar became very

33

Oriental windmills employed horizontal sails made of cloth and bamboo.

inquisitive about it, although the principle did not agree with his religious beliefs. He searched in foreign books for information and finally encountered what he wished to know in an Italian volume brought to the Orient. It contained crude sketches of a windmill, which Wang Cheng copied. Seven years later he published a book titled *Strange Machines of the West*. It was so huge it had to be divided into 750 volumes. In it was detailed information about the windmill. But although the Chinese used wind power for pumping, they never employed it in any other manner. Preferring not to anger their gods, they went on doing their work by hand in the old laborious way.

Finland was endowed with plenty of wind, and in its Åland Islands, nine hundred post mills at one time ground the farmers' grain. Finland today has great interest in developing wind power.

Windmills were especially popular on islands. About ten thousand small windmills were still pumping water in Greece and its islands in 1961, though since then electricity has taken over. Some of the mills had as many as a dozen cloth sails. The island of St. Helena, off the African coast, where Napoleon was exiled, had two windmills to do the threshing. In the Azores, customers were summoned to the

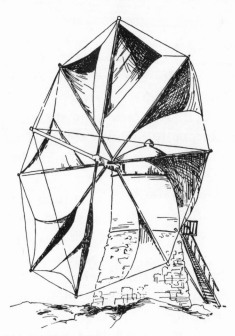

Mark Twain said the windmills of the Azores were so lazy they ground only ten bushels of grain each day. When the wind changed, donkeys were hitched to the beam, and the whole upper half of the building was turned on its high stone base.

mills by the blowing of a conch shell when it was time to bring their grain. The tallest building in the Channel Islands was the windmill at Sark, built in 1571 under a charter from Queen Elizabeth.

In Holland, a man's wealth was measured by how many windmills he owned, so naturally Dutch settlers took their tall towers to Java, South Africa, and other Netherlands colonies hundreds of years ago.

Christopher Columbus, on his third voyage, brought a millwright, grinding stones, and everything necessary to build the first mill in the New World. It was expected that Spanish colonists would want to use the kind of machines they had at home. As it happened, crushing grain for flour did not become the principal application of mills in the West Indies. Sugar cane became the main crop in the islands, and the mills were soon adapted to this new industry. In the

beginning the cane was crushed between three upright rollers, the middle one turned by sweeps to which horses or oxen were harnessed. But farm animals were exceedingly scarce, and, so the early Spanish colonists looked to the wind to help them. In places like Barbados the trade winds blew steadily a large part of the year.

The West Indies were owned by several nations, among them Denmark and Great Britain, whose "Sugar Islands" sprouted windmills by the dozens. Those solid white towers still standing are great tourist attractions.

On the mainland of North America, the first viceroy of Mexico was encouraging his Spanish followers to build windmills for grinding grain. Whatever structures they erected for this purpose have long ago disappeared. In Mexico, the grinding stone of the aborigines survived longest. Only in recent years have electrically driven mills taken their place in furnishing cooks with the ground corn for tortilla paste. However, Mérida, in Yucatan, became a city of modern pumping windmills to supply water for the populace.

Windmills entered the United States with the first colonists. This was also true in Canada, where the French brought windmills to Quebec. Sir George Yeardley, governor of Virginia, erected America's initial windmill in 1621, and several more soon sprang up. (A post mill has been reconstructed at Williamsburg, the old site of one.) At New York City the Dutch raised two windmills in 1622, and farther north the Pilgrims built the first New England windmill at Watertown, Massachusetts.

Each village of any size hastened to erect either a wind- or a water mill, for flour was needed for the growing settlements. At first, horse-driven stones served in Manhattan, but as quickly as materials could be brought, a post mill reared its sails above the Battery. The windmill belonged to

In New York City in 1859, Manhattan's only "skyscraper" was the windmill behind the fort at the Battery.

the Dutch West India Company, for this was in the time when feudal overlords claimed to own the wind and only the Crown company had the right to use the breeze.

Indians were startled at the sight of the mill and believed an evil spirit drove its great arms. This modest mill was New York's first skyscraper and its first traffic signal. The Brooklyn ferryman used it as a weather vane, for he knew that if the sails were furled there was likely to be a storm, in which case he refused to take passengers across the East River.

Abraham Pietersen, a grouchy miller from Haarlem, Holland, crossed the Atlantic to operate the Battery windmill. He was constantly being hailed into court because of his bad temper. One judge, tired of his constant quarreling, ordered Pietersen to hold his tongue and tend to his milling.

The Battery mill was used until 1661, but by the time the Dutch surrendered to the English three years later, its work was being done farther up the island and the old building was falling into ruin. There used to be a little hill called Catiemuts (or Katy's Bonnet) near where the Woolworth Building now stands, and here the Garrison windmill was built. The miller ground twenty-five "skepfuls" of flour for the militia each week to pay for using the wind. Although this mill was destroyed by lightning, another was erected on the spot and stood there until a street was extended from Broadway in 1740.

Cortlandt Street was called Old Windmill Lane because of the mill that stood there. Two more were erected farther up the island to grind snuff and run a brewery. Such mills were so closely associated with the character of the city that its official coat of arms showed the crossed sails of a windmill and a flour barrel, indicating that when James II granted its charter, New York was considered a milling town.

chapter 8

New England's Windmill Belt

Settlers on the easternmost end of Long Island pumped bogs, drained land, and ground grist with their English-style smock windmills, which had rotating caps on the towers. Among the Kentish folk who founded the village of East Hampton in 1649 was a noted family of windmill builders. They erected two towers, one at each end of the wide main street, which has survived to modern times. In fact, this is one of the few places in the United States where an old-style windmill can still be seen.

Often lack of a breeze left the village without flour. During a long spell of calm weather one of the millers went to visit friends at Montauk Point. In the middle of the night he was awakened by the sound of a rising wind. He hurried into his clothes, saddled his horse, and rode home to start his mill as quickly as possible. Leaving his horse tied to the fence, he worked the rest of the night, stopping only when the breeze died in the morning.

On another occasion the villagers were attending church when the minister noticed that a long-awaited breeze had sprung up.

"Church will be dismissed," he announced. "Miller, go to your mill and grind." The minister was thinking partly of his own benefit, for according to law, the pastor's grain had to be ground first.

Early in the nineteenth century Sag Harbor had a windmill that could be seen from a great distance, since it stood on a hill. Each time a whaling vessel returned to port, a flag was hoisted on the tower. This brought relatives from all over, hurrying to greet the returned voyagers. The Sag Harbor mill was moved several times. Twelve oxen were required to drag it to its present site in the village of Water Mill, where it ground corn and wheat for many years.

Though it seems incredible today, the United States once had a windmill belt as picturesque as those in England and the Netherlands. Beginning on Long Island, it extended through Rhode Island and out to Cape Cod and Nantucket.

In these oldest mills scarcely a nail was used, for spikes of any kind were precious articles among the early colonists. Strap iron bands to take the strain off the beams were forged by village blacksmiths. Bolts were often held together with a piece of horseshoe passed through the eye. Wooden pegs and cogwheels made of tough white oak were patiently fashioned by hand. The millstones, usually six feet across and over a foot thick, were brought from England. When they needed dressing—that is, having their furrows refinished—they were lifted out by means of homemade equipment resembling traveling cranes. To dress any millstone the surface had to be painstakingly gouged out in such a manner as to spread the grain evenly over the stone. One in good condition did more than crush the kernels; it split and spread them out and scraped the inner part free of bran. Some New England mills also had corn crushers, a unique invention.

Those old windmills were no newer in design than the

ones the earliest colonists had seen in their native lands. Long Island had a mill patterned after the most ancient sunken post mills. At Newport there was a "spider mill" with jib sails like those on the stone towers of Mediterranean countries. Probably a returning sea captain or a Portuguese whaler who knew such mills from his native land had taken part in planning it.

The Pilgrims built smock mills with eight straight sides that reached down to the ground like a skirt, reminiscent of the smocks that French wagoners of that day wore. France had some mills the same shape. The long beam turned the whole top of the tower and made a good brace in stormy weather, when it was chained down to hold the building steady. When the wind veered, the beam was pushed around on its iron wheel.

For many years every high spot on Cape Cod had a windmill, and these served very well as landfalls for seamen. While some ground grain, others operated salt works, producing hundreds of thousands of bushels annually. Long rows of vats held the water the towers pumped into them. When the weather was fair the lids were taken off the tubs and the breeze evaporated the moisture, leaving a deposit of salt. Until 1834 almost all the salt consumed in the United States came from Cape Cod. In a single year 600,000 bushels of salt were produced with the aid of windmills. The industry was gradually abandoned when the duty on foreign salt was reduced and America could buy this commodity more cheaply abroad.

Back in 1633 a sturdy wooden mill was erected at West Yarmouth, and in 1894 it was moved to Chatham, where the owners kept it in good repair and set the stones grinding occasionally so as to make sure the machinery was still in order. The old tower required a lot of care, since the sails had to be mended frequently, and every spring vines had to

Nantucket mill was built in 1746. It has eight sides, like all smock mills, and a peaked bonnet.

be cleared off to prevent decay. But the owners knew it was a treasure and treated it accordingly. By then it was the oldest surviving mill in the United States, and some friends of Henry Ford, knowing that he admired early American buildings, came to Massachusetts and bought it for him as a birthday present. Despite loud protests from the townspeople, the tower was carefully dismantled and moved to Dearborn, Michigan, where it was set up in a museum.

Nantucket Island has this country's most interesting windmill, constructed partially of timbers from wrecked ships. It was built in 1746 to grind corn, saw wood, churn butter, and sharpen whaling weapons. When the British captured Nantucket during the Revolutionary War, the loyal islanders set the sails of the windmill as a signal for their ships at sea not to come in. The British waited for them in vain; none of the Nantucket boats returned home, and the mill had saved them from falling into a trap. This mill was still grinding grain as late as 1974.

In a park at Newport, Rhode Island, stands a round stone

What is left of Governor Arnold's
mill at Newport.

The Chesterton mill in Warwickshire,
England, from which he copied it.

structure that aroused the curiosity of the poet Henry
Wadsworth Longfellow. He was told that a Viking was
supposed to have erected it for his ladylove. Longfellow
wove the story into his poem "The Skeleton in Armor,"
which includes these four lines:

> *There for my lady's bower,*
> *Built I the lofty tower,*
> *Which to this very hour*
> *Stands looking seaward.*

Not everyone agreed with his theory, pointing out that if
the Vikings had left the building there, why had not the first
English colonists mentioned it in their records? The tower
remained a mystery until it was recalled that the first
Benedict Arnold, who became colonial governor of Rhode
Island, had written of living near such a building in England
in his boyhood. It was the Chesterton windmill, designed by
the famous London architect Inigo Jones, who did not

believe in imitating the Dutch. Instead of a conventional mill, he put six stone columns around the base of the Chesterton mill and covered the roof with a flat lead dome.

The structure at Newport had the same kind of arches, so the authorities reached the conclusion that no Viking had anything to do with it. The mystery was finally solved through Governor Arnold's own writings.

When Newport's first wooden mill blew down in 1675, Arnold decided to have it rebuilt in such a manner that it would stand through storm and fire and look to the Indians like a strong fort. He took the Chesterton mill as a model, and his stout walls still are there, though the roof and wings are long since gone.

chapter 9

Whirling Sails and Brave Millers

By the early part of the eighteenth century, old-style windmills had reached their greatest importance and were used to perform a variety of essential tasks. A single windmill could raise ten thousand gallons of water four feet in one minute. In a moderate wind it could grind five hundred pounds of grain in an hour.

No one could foresee that the gallant army of four-armed giants would soon be doomed. A new rival was about to come on the scene at the richest period of their career. But before that happened, the world witnessed an age of great mills and brave millers and their dauntless achievements. Even the skyline of London was filled with spinning sails, and by 1750 there was a windmill to every square mile of the city. That is why there are such places there as Milford Lane, Mill Wall, and Windmill Street.

Long ago windmills pumped water to drain the marshy land upon which London was built. Later, after the Dutch and Norwegians had harnessed the wind to saws, the British brought workmen from Holland to erect a wind-driven sawmill at Limehouse to cut ship timbers. English workmen

Repairing sails on the big old-style mills was a dangerous task because the work had to be done high above the ground.

watched the preparations with alarm and, thinking they were about to be deprived of their jobs, rushed at the new structure with axes and drove the Dutch away. The building was never completed.

The very height of a windmill, the way it loomed above everything else for miles around, seemed to give the miller importance in the community. Moreover, a man who could climb so high on dark nights when gales blew deserved to be regarded as a village hero. His life might be a peaceful one of gossiping and grinding until he sensed a storm brewing. Before improvements were made in the handling of the sails, this meant he must take them in by hand, a very difficult task. The cloth had to be drawn from each arm, and afterward the miller had to climb the framework to fix the

The high circular platform on this mill in Rotterdam was intended to make maintenance of the sails easier for the miller.   *Photo by Werner Lenggenhager*

fabric in place again. Many a man lost his life doing this, for there was a chance of his being sent spinning to the ground.

One day in Holland an old miller was repairing his sails while a boy helper held a brake to steady them. A passing friend called to the miller and he answered from his position on the tip of the sweep. Suddenly the blades began to whirl and the astonished miller hung on for dear life, looping the loop. The neighbor rushed inside, stopped the mill, and asked the boy if he was trying to kill his master.

"No, sir," the boy said. "I heard him call and thought he was on the ground, so I let go of the brake."

The miller looked at his gnarled hands when he reached

safety on the ground and observed, "You may be old, but you're still worth a lot."

For centuries no one knew a better way to handle the sails than stretching them over a lattice on each blade and furling them to catch exactly the right amount of wind. The farther they were unrolled, the faster the stones went.

Often a balcony was built around the tower for the miller to walk on while he tended the sails. One night a miller, while turning his blades into the wind, stumbled over a stick left from some repairs. Down came the whirling sails and the miller was beaten to death. His helper came out on the balcony to look for him, tripped over the same stick, and fell to the ground.

After accidents like these, inventors went to work on making the sails easier to control. Among the first was John Smeaton, who received a gold medal in 1759 for his improvements on windmills and water mills. There were others, like Stephen Hooper, who designed roller blinds with remote control, and William Cubitt, who combined Andrew Meikle's shutter sails with Hooper's controls, eliminating the need to roll up the sails.

In old-fashioned mills, which couldn't turn quickly out of the wind, the great fear of the miller was that he might run out of grain and be unable to stop his sails in a storm. As long as there was something to grind between the stones, the machinery could race along without any damage. But the moment the bins were empty the stones threw off sparks, and then the shaft could catch fire and send the whole building up in flames.

Jeremiah Shaw once owned the "best blown" mill on the banks of England's Mersey River. On a Sunday night in 1839 Jeremiah was alone when a terrible storm blew up and the sails got out of control. The miller opened his hoppers and fed in every sack of grain in the place. When it was used up, the shaft and the wheels tore around, getting hotter by the minute. Then Jeremiah tried the last remedy he could think of. The mill had a beam which he sometimes used as a brake, but it never held very well. Resigning himself to a bad night, the miller braced the beam against the shaft, then sat down on the big stick, bouncing up and down like a jumping jack. With his added weight, the timber slowed the shaft enough to keep a fire from starting. Jeremiah was not as fortunate as his tower; by morning he was a sick man and never recovered from the terrible shaking.

Tall grain mills were very different from the peg-and-post push-around houses. The shaft still rose through the center, with cogwheels to turn the machinery. There was a floor for the meal bins and the wheels for adjusting the grinding stones. A shaking tray, called the "miller's damsel," regulated the flow of grain out of the funnels and into the hole in the stones below. If the wheat ran out of the hopper too fast, the miller might shout to his helper, "Don't choke her eye."

Everything about an English mill was generally spoken of as feminine. The outside of the stone was the skirt, the center part was the waist, and the grinding surface required dressing. Other words in the windmill vocabulary were scoops, scuppits, spudges, bushels, fag-hooks, sack boys, tedders, pritchels, skeps, and sieves. The sails had three parts—canister, middlings, and sweeps. Circular wooden shields, or vats, enclosed the stones to keep the flour from escaping.

Until 150 years ago, millers did not bother to sift flour, leaving this task to the bakers. When sifting did finally become part of the miller's work, he employed a cylinder-shaped sleeve of bolting silk, a mesh through which the fine powder passed at the lower end while the bran was shaken out above.

In front of the door of the mill, customers tethered their horses and came inside to wait for their grain to be ground, or whatever the mill produced. In any weather the mill was always warm and cozy, a pleasant place for neighbors to enjoy a chat.

The great weakness of windmills, which became more important as the years passed, was that when the breeze died they were idle. A brisk wind might generate 50 horsepower, but when it dropped to eight miles an hour it barely turned the stones. Then the miller might have to wait a week for power to continue his grinding.

In Liverpool one summer there was neither wind nor rain for many weeks. Water mills and windmills were completely idle, and the supply of flour was running very low in the city. Two bakers had plenty of wheat waiting in a certain mill and at last, late Saturday night, they observed that the wind was rising. They rushed out to awaken Old Richard, the miller, and tell him to get busy while there was a breeze.

Liverpool had a law forbidding work on Sunday, and Old Richard told the bakers, "I'm not going to pay a fine and I can't hide my sails."

Each baker offered to pay the cash penalty, and they got into a big argument. Finally the miller reminded them that the wind might die suddenly. Each wanted his grain ground first, and it was dawn before a solution was found. Finally one baker offered, "Let Master Richard take first a sack from your pile and then one from mine. We can divide the flour and the fine evenly."

"Very well," the miller agreed. So all day long his sails whirled in full sight of the churches across the way. When Monday came the wind died again and the two bakers were the only merchants in the city with bread to sell.

chapter 10

Language
of the Sails

In Holland lived two brothers, who both owned windmills. Though some distance apart, they were within sight of each other's sails. To visit back and forth required a walk of several hours, and when their aged father lay at the point of death, the elder brother had to find a means of keeping the other informed. Telephones had not been invented yet.

"If Father gets worse," the older brother proposed, "I will take the boards off my sails. If you see that they are gone, come at once."

This practical plan was the beginning of the windmill telegraph, and it soon became the custom in the lowlands to convey various messages by removing boards from the wings or manipulating the canvas sails. If one understood the language, there were ways to warn of weak dikes and coming floods and to invite one's neighbor to dinner. If the miller himself died, the sails stood idle; all twenty crossboards were removed, and as part of the funeral the arms were turned slowly to follow the body to its last resting place.

For good news, decorations were added to the arms. A

In olden days a mill like this was said to be "standing pretty." It was decorated for the marriage of the miller's daughter and the wedding guests would dance on the balcony.

wedding called for tin hearts, flags, and garlands. The English had another code, signaling by the position of their sails whether the mill was grinding or was stopped for repairs.

The business of smuggling sometimes was aided by less scrupulous millers. On the Sussex shore contraband brandy, tea, tobacco, and silks, brought in small boats from France and Flanders, often were hidden in the towers. By setting the sails in a certain manner the miller signaled when it was safe for boatmen to land their cargoes. The goods were carried up in fishermen's nets and bulging boat bumpers.

Near Brighton lived a miller named Lot Elphick, who owned two boats whose crews he kept informed by the sails of his mill. When he was not on the ground, his foreman set the sails to give him the latest news. Lot might be visiting in town, but from the handiest window he could look at the wings on his tall tower and tell what was going on there.

One day he was at the vicar's house talking about a donation for the church when a company of dragoons rode by. In the distance he noticed his sails were whirling a frantic signal to him. He realized that the soldiers had learned of a thousand pounds' worth of brandy stored in the mill and were riding there to make arrests. Lot took a chance and told the vicar of his predicament.

The miller had been generous to the church, and the vicar did not wish to see him go to prison. "I will help you," he offered, "if you will ask me no questions."

Elphick promised, and together the pair mounted horses. Instead of heading for the mill, they went straight into town and the miller realized they were going to police headquarters.

"You are betraying me," he accused the vicar.

"You promised to leave everything to me," the churchman reminded him.

So the miller acceeded and listened nervously when the vicar said to the commissioner, "My friend, Lot Elphick, came to me today to ask my advice. He wanted to know what to do about the smugglers who insist on using his mill. They have been troubling him a long time, but now you have saved him further annoyance and he wishes to thank you for sending troops to clear the mill of smuggled goods."

Lot's jaw nearly dropped off with surprise as the commissioner shook hands with him. By the time the soldiers were back, Lot had promised the garrison a present of the brandy casks. No one knew that, had it not been for the sails of the mill, he would have gone to prison.

The practice of employing windmills to convey messages lasted into modern times. The Dutch made good use of their sails during the Second World War, when the Germans took the country. The Underground was alerted to enemy plans by this means. Allied pilots whose planes had crashed were given help out of Holland when their presence was revealed by the language of the windmills.

chapter 11

Strange Shapes
and Strange Names

Since man learned how to harness the wind, he has never ceased to experiment with it. A South Carolina railroad once had a wind-driven locomotive, but train engineers did not know what to do with the sails when the breeze changed. In England a man patented a kite carriage in 1826, using the same principle, but his invention frightened people by its speed—twenty miles an hour. During the early seventies, a California firm tried building windmobiles as a substitute for automobiles.

Windmills have been placed on ships, and in the eighteenth century a French military expert proposed invading England by means of floating forts built on huge barges. Each was to be propelled by four paddle wheels driven by an equal number of windmills.

A man in Florida built a pleasure boat driven by two seagoing windmills, storing compressed air in tanks. As the air was released, it provided energy for a small motor. But there were drawbacks, for the engine made a terrible whistling noise.

Sometimes the windmills themselves were given fanciful

The Battersea windmill had walls made of shutters that could be opened and closed. Inside was a huge upright paddle wheel driven by the wind.

shapes in the past. A monstrous one erected on an estate at Battersea, England had its sails concealed inside a wall made of ninety-six shutters, which opened like Venetian blinds when a rope was pulled. They stood eighty feet high and let in enough wind to drive a mighty paddle wheel made of planks set up around a central shaft.

This odd-looking structure was 54 feet across and 140 feet tall, overshadowing the nearby village church. Because of its resemblance to a gigantic packing case, a curious story circulated about the mill: The emperor of Russia, while visiting England, took a fancy to the Battersea church and decided to carry it off, so he had this large container built for it. But the inhabitants, the tale went, refused to let their church go, and the tall packing case remained where it was put. The mill inside ground linseed for oil and later malt for a distillery.

Modern technicians have revived the upright design for windmills, but they do not bother to build a tower for them; they are set out in the open.

The old-type mills were often given unusual names for some reason important to their owners. For instance, Peter the Great of Russia in his youth spent some time in Holland working for a shipbuilder. A new windmill went up nearby,

The windmill at Potsdam near the palace of Sans Souci. This is the one Frederick the Great was unable to buy.

and Peter was designated to nail some boards on it. When the owner heard of this he named the mill the Czar of Muscovy.

It was the custom to put carvings, paintings, and a nameboard on some of the mills, just as sailing vessels had figureheads. Some of the names were very strange, such as Honey Vat, White Death, School Master, Constant Battle, and Pelican.

One of the better known mills in Europe was that of Frederick the Great at Potsdam. When the king built the palace of Sans Souci he decided to add a windmill that stood near the royal park. He sent his agent with money, but the owner refused to sell at any price.

"Then the king will take it and you will get nothing," the agent threatened.

"Not even the king can do that while we have courts of justice in Berlin," the miller answered.

When the agent reported to the king and suggested that the miller should be punished for his insolence, Frederick laughed, pleased with the man's faith in the country's laws. Instead of turning him out, Frederick permitted the miller to remain on the palace grounds, where his tower stood until modern times.

chapter **12**

Wind Meets Steam

With the coming of the nineteenth century, something happened that spelled doom for many thousands of windmills the world over. The steam engine, patented by James Watt in 1769, was applied to milling. Until then, bread everywhere had been made from soft wheat, which could be crushed easily between stones. People did not mind some grit in it and pure-white flour was unknown.

It all started in Hungary. On that country's plains only a hard wheat, unsuited to the motion of stones driven by wind or water, could grow. Because of this, the Hungarians began experimenting with the then little-known steam power, and in 1820 they built the first steam-driven flour mill. Twenty years later a man named Papst used the first iron rollers for grinding grain. His flour was fine and clean, but the millers were slow to change, regarding the new invention with suspicion.

Just as the idea of the windmill had traveled from one country to another, so it was now with the steam engine. There were many other countries which, like Hungary, could grow only hard wheat, and so needed the steam-driven rollers. In the United States and Canada the new method was especially welcome, and an altogether different type of flour mill sprang up in the big cities of the North American wheat belt. More slowly, Europe followed suit, discovering that steam could be applied to many of the uses for which the leisurely windmills had formerly served, and there was no waiting for a breeze to get a task done.

As a consequence, towers fell into disrepair and sails went unmended. Field mice moved in as millers moved out. Domes leaked, machinery rusted, ladders were broken and floors rotted. Like so many old men, the windmills in much of Europe were tottering, and only old men were loyal to those that were left.

France printed a postage stamp that celebrated such an old windmill and such an old man. Alphonse Daudet was an author who liked the winged towers so much that he bought one for himself in the Rhone Valley. It was built of stones and covered with moss and vines. An owl lived in the turret, and rabbits were quartered where grain sacks had once been stored. The stones had been idle for twenty years. Daudet wrote about the mill's former owner, the Miller Cornille, in a book called *Letters from My Mill*.

In Cornille's youth almost every little hill in that part of the Rhone Valley had windmill sails whirling above the pine trees. Paths leading up the slopes were alive with donkeys carrying sacks of grain or flour. When the mills rested on Sundays, villagers made merry with the miller and his family. Country wine was served and a piper played for dancing.

All this ended when men from Paris erected a steam-rolling mill on the road to Tarascon. The donkeys with their grain sacks began traveling in that direction, and one after another the windmills ceased operating. Their sails stood still, and grass grew long in the paths through the pine trees.

That did not happen with Miller Cornille's sails. The old man, already past sixty, proudly boasted that "God's own breath" still turned his mill and he had grain to grind. No one listened to his chatter, and he became offended by his neighbors' indifference and shut himself in his mill, not even permitting his granddaughter, Vivette, to enter. Her parents were dead and, although she was only fifteen years old, she had to work in the olive groves, or in the harvest, or tend silkworms to earn her living. Neighbors were appalled that the old man did nothing to help the girl.

Their anger at Cornille gave way to curiosity about him, for they saw him on the road every night, his mule loaded with full flour sacks. They wondered where he obtained his grain and how he kept his windmill so busy. When they asked, he said the flour was to be shipped away for export.

There was a continual mystery about the turning sails, the mule grazing nearby, and the old man hiding inside the tower.

After a while Vivette fell in love with the eldest son of the village piper, who had once been Cornille's best friend. The boy's father went to ask the miller's approval of the marriage, but Cornille refused to unlock his door.

"If you want your son to marry, find a wife for him among the girls at the steam mill," he shouted.

Vivette and her sweetheart decided to talk to the old man themselves. They arrived at the mill one evening after he had gone on his nightly errand with the donkey. The door was closed, but Vivette climbed a ladder and looked through a window. In the fading light she saw that the interior was empty. There was not a sack of grain or flour, and the drive shaft was covered with dust, appearing as though it had not been used in a long time.

The young couple climbed down inside and searched the tower. They found a few tattered clothes, a rumpled bed, a little food, and in a corner three or four whitish sacks filled with chalk. Now they knew the miller's secret and that his trips with the mule were make-believe. He had no wheat, no work, and no money.

The two young people told the piper how poor the miller was, and he appealed to neighboring farmers to bring their grain as quickly as possible to the windmill. The whole village was soon on the road with a procession of laden donkeys. When they reached the windmill, Cornille was sitting on the stone step and the door was wide open; he knew his secret had been discovered. "There is nothing left for me to do but die," he moaned. "The mill is dishonored."

At that moment he saw the laden donkeys. Scarcely believing the miracle that they were bringing him grain, he watched the villagers as they piled their sacks beside the door.

"I knew someday you would return to me," he said. "The millers at Tarascon are thieves."

He felt his family honor had been saved and gave his consent for Vivette to marry. The villagers, pleased with his happiness, invited the miller to town to celebrate.

"No, no, my children," he responded. "First I must give my mill something to eat. It hasn't had anything between its teeth for a long time."

The villagers continued to bring him their grain, and Cornille's mill was the last to grind in the valley.

By the beginning of the twentieth century, only 2,500 windmills were left in Holland. In the fields they had pumped dry, their places were taken by steam pumping stations. Between 1923 and 1927, the drainage system for one farming region lost fifty handsome mills that had been in use three hundred years. About the same time a campaign began to save the others from demolition and neglect, but it did not succeed very well because it was considered old-fashioned not to install modern pumping stations. Oil and electricity came to the rural districts and took over more of the work of the windmills.

Then along came the Second World War, which devastated even more mills, leaving only about a thousand. But there were people who had a great affection for these landmarks and were dismayed by what was happening to them. The Dutch Windmill Society and the Guild of Volunteer Millers were formed, and they succeeded in having laws passed to aid in the preservation and restoration of the towers. Regulations prohibited altering or tearing down any windmill without special permission by the provincial authorities. Some of them were turned over to municipalities, which today support the upkeep of the towers as public monuments.

As a consequence of this campaign, Holland is still a place to see old-fashioned windmills, even though they no longer pump water or grind flour. Their main purpose today is to beautify the landscape, except for a few that serve as museums or, in some cases, as homes for families who like the picturesque.

England's remaining mills are cared for by the Society for the Protection of Ancient Buildings. The North Leverton mill at Nottingham, supported by gifts and loans, keeps its stones turning and grinds flour, which is sold to tourists to help finance preservation.

In 1958, only twenty windmills were still in use in Great Britain, but today more than fifty wind or water wheels are grinding grain, and others have been saved and restored to become educational centers or tourist attractions. A mill near Croydon is on the grounds of a school, and the students use it as their dressing room for sports. Other mills are open to the public, among them one of the very oldest at Bourn, near Cambridge. It dates back to 1636 and was last worked in 1927. England found an unexpected use for its ancient mills in 1949, when there was a record grain crop and thirty of them were put to work grinding.

In West Germany, people who have acted to preserve windmills were publicly thanked by having their names listed in the Honor Book of German Mills. Preservation societies exist in other places as well, such as the Friends of Windmills in Cartagena, Spain. The United States has windmill preservation societies on Long Island and in New England. All of the old towers remaining have become public charges, and most of them are being shown to visitors.

chapter 13

Invention of American Windmills

At a time when European windmills were beginning to decline, with some being sacrificed so their sites could be turned to commercial use, a different story was unfolding in the United States. One of those who felt it was a shame that this free, clean source of power should be abandoned was Daniel Halladay, who in 1854 invented the American version of the windmill. He realized that some kinds of factory work might be better done by steam, but there was one chore at which the windmill excelled; it was the best and cheapest pumper a farmer could have. Halladay saw no need for tall wooden or stone towers, nor for a great amount of wind. The day was past when such structures could be assembled by village carpenters; for a farmer in the nineteenth century they would be hopelessly expensive.

In the Middle West and the Southwest settlers and ranchers had little money or material with which to bring water to their dry acreage. The cattle-raising industry spread into areas where there was plenty of nutritious grass for grazing but surface water was scarce. Following a rain the plains might be dotted with small lakes and water holes,

Some Nebraska farmers used to erect jumbo mills consisting of a paddle wheel set in a box. The protruding blades were pushed around by the wind.

but they dried up after a few days of sunshine and wind. Any place with a constant water supply was settled to capacity.

Seeking power for pumps, early ranchers gathered short lengths of wood left over from building cabins and barns and managed to assemble grotesque windmills with fan-shaped paddles for arms. They nicknamed these machines go-devils, merry-go-rounds, or battle-axes. Some constructed paddle wheels set on the ground, with one blade at a time showing over the top of a large box. All of these monstrosities were cheap to make, but they were not very efficient. Halladay believed he could create something neater, more effective, and small enough to ship anywhere. His finished product was a round fan composed of tiny wooden slats, with a tail to set it whirling in the breeze.

In Haverhill, England, there was a unique windmill, which may have been Halladay's inspiration. Its eight arms, instead of holding eight sails, supported a continuous circle of vanes, each five feet long and a foot wide.

Halladay's creation was much smaller, but it caught on in no time at all and within a few years these modest fans were seen on top of clapboard towers from Texas to California and all across the Great Plains. They pumped water into cattle troughs, irrigated gardens, and stood alongside railroad tracks to refill the tanks of locomotives.

The early Halladay windmill was often erected on top of a wooden tower that contained a tank.

In 1883, Thomas O. Perry gave this truly native American windmill shining steel vanes instead of wooden ones and placed the mechanism atop a slender metal tower. A furl wire at the base enabled a farmer to set the blades into the wind. No other control was needed.

Unlike the big European windmills, the American ones could be shipped by train or wagon. They could pump water from any depth. And although they did not pump a large volume of water in a short time, as did the irrigation pump, they provided amply for cattle and at the same time did not deplete the underground water supply.

Ranches in the Middle West might have ten or fifteen mills; a few of the largest had several hundred. They were checked once or twice a week. Sometimes cowboys took care of the maintenance and their primitive lubrication systems, other times an expert from the nearest town was employed. The local windmill man was an important figure in prairie country.

On a hot August day a cow might drink twenty-five gallons of water. If fifty cows watered at the same tank, their owner would have to turn on an electric pump to help out, or he might have to haul water in his pickup until the windmill began turning again.

American mills performed so well that their basic design underwent only slight changes in the course of a hundred years. In the 1800's something like 25 percent of the United States' nontransportation energy was supplied by them.

As a rule these little mills did not inspire the kind of stories that emanated from the old Dutch ones, but here is an anecdote dating back to 1908. One day during a gale a man ascended a ranch tower to remove two broken sails. He was obliged to stand on the outer circumference of the wheel while he worked. The wind carried one of the spokes out of his hand, it dropped and struck the pin holding the furling lever in place. The pin was knocked out, the lever fell, and the sails immediately turned into the wind and began to revolve. The man, hanging on by hands and feet, was carried around several revolutions before his shouts were heard by another fellow some distance away, who came running and stopped the wheel just in time before the repairman became exhausted and fell.

The American towers were rugged. They could operate unattended, occupied little space, could be erected easily, and cost relatively little. Soon there was a demand for them in other countries, too, and in Argentina they literally made history. Very few streams water the eastern slope of the Andes Mountains and, except for the branches of the Río de la Plata, many rivers often went dry or were lost in shallow, brackish lagoons.

Argentina had always been a great meat-producing nation, and many of its men practically lived in the saddle,

The old and modern ways of watering cattle in Argentina. Instead of long buckets made of hides and pulled out of a well by a man on horseback, steel fans on high spindly legs pump water into tanks.

tending herds and driving them long distances from one watering place to another. In early times there was no way to quench the thirst of livestock other than with leather buckets pulled up from wells. Some years the pampas region was dry as bone, and the water holes were choked with dead steers, their bodies trodden many feet deep in mud by those that came later.

If the inhabitants knew there were such things as windmills in other parts of the world, they did not construct them because in those endless plains they had neither wood nor stone with which to build them. Wind was there aplenty, but not the means of harnessing it. Therefore the invention of the small circular fans on top of metal towers was a great boon to Argentina. Boatloads of them were brought in, new wells were dug, and the wind began pumping water into cattle troughs. Estates sprang up around the watering places, and the silver fans supplied the precious liquid for ranch houses and gardens. The pampas lost that forbidding look and became dotted with steel towers. Small towns became veritable forests of windmills, with one in nearly every yard. The railways installed them alongside the tracks, and a traveler could look ahead and see the towers rising from under the horizon as his train moved along.

Similar transformations took place elsewhere. Experts

regarded the mills as an important step in colonizing the great open spaces of Australia. Although the Dutch-type windmill became obsolete because it was costly to build and maintain, the multivane fan was accepted the world over, and companies in several countries began manufacturing and exporting them.

Far out in the North Pacific Ocean the slender steel towers changed barren, uninhabited islands into attractive gardens. Before the era of jet travel, refueling stations were required, and engineers mapping the route of clipper planes to the Orient selected five islands as landing stations. These had to be agreeable places where passengers could rest while supplies were taken aboard. Above all, plenty of fresh water was necessary, therefore steel windmills were among the first structures erected at each station. They pumped water into storage tanks and irrigated the ground as well.

One of these stations was Wake Island, a dismal, lonely spot with little vegetation. Clever horticulturists, once ensured of sufficient water pumped by windmills, converted Wake into a pleasant place to live in.

chapter **14**

The Coming of Electricity

Late in the nineteenth century a new and unforeseen use for the wind unfolded. Engineers had discovered how to generate electricity with steam plants, and a Danish professor, P. LaCour, in 1892 found a way to produce this important new kind of energy with the rotating shaft of a windmill. He used a regular four-sailed Dutch-type mill at Askov, Jutland. Being an extremely windy region, it is an excellent place for experiments of this kind.

News of LaCour's success reached the Norwegian explorer Fridtjof Nansen, and a year later, when Nansen sailed for the Arctic to search for the North Pole, he took along a generator and a small four-winged windmill. Although the equipment must have been extremely crude, it worked well, and when the ship *Fram* was icebound and drifting in the polar seas, it had electric lights at a time when New York and London were still illuminated with kerosene and gas.

The polar regions were so windy that Robert Scott's British Antarctic expedition set out in 1900 similarly equipped to generate electricity. Machines were considerably improved by the 1930's, when Rear Admiral Richard E.

Byrd also headed for the Antarctic. He carried along a Jacobs wind generator and left it there. His son went back twenty-two years later and reported that the windmill was still at work and in good condition, although deep banks of snow had caused the tower to lean. He dismantled the apparatus and took it home from Little America, proud that it had been standing so long in winds that sometimes exceeded a hundred miles an hour. A prediction was made that someday the South Pole would be surrounded with gigantic windmills for the purpose of producing energy for industries.

Professor LaCour's success led to the formation of a Danish Wind Electricity Society, and by 1908 seventy-two electricity-producing wind machines had been built in his country. The number reached 120 in 1918, spurred by the need for power during the First World War.

It was that same conflict which stimulated interest in America in uses for propellers and lightweight metals. Experimenting began in several widely separated places, and some of it may have been inspired by former members of the Air Force, who mounted propellers on top of their houses to provide power for radios. Two Iowa farm boys, John and Gerhard Albers, in the 1920's became annoyed at having to take their battery-powered radio set to be recharged at a garage once a month. They didn't like to miss programs while this was being done, and their father begrudged them the hard-earned dollar it cost every time.

One day Gerhard happened to read in Nansen's book *Farthest North* how his expedition had enjoyed the advantages of wind-powered electricity while drifting in the polar ice. There was a picture showing the windmill rigged on deck.

Gerhard was excited by the idea and immediately went to work on his own version. With an abandoned metal

Tiny two-bladed windmills like this used to be seen by the thousands on rooftops across the United States before rural electrification became commonplace. They provided energy for radio batteries and farm equipment.

windmill and tower, an automobile generator from a junk yard, and gears from a cream separator, he achieved similar results. However, he and his brother thought they could improve upon this with propeller blades whittled out of wood. They tried out the idea by attaching the blades to the family automobile and driving it along a road to simulate the force of the wind. By trial and error they designed a propeller capable of supplying light and power for the farm.

The brothers believed they could produce the small windmills commercially and placed an advertisement in Iowa farm papers. Orders came in faster than they could keep up with them, so they took their savings and assembled a machine to do the whittling they had done by hand. They ran it with power from a tractor outside their workshop window. Their next step was to find financial backing so they could rent a factory in Sioux City and begin making and selling the entire rig—propeller, generator, and tower. This was the stage they had reached when an engineer from a large radio manufacturer contacted them. He had been looking for a means of adapting windmills to radio receivers on farms remote from power lines. At that time less than 20 percent of the farms in the United States were electrified. The Albers brothers agreed to merge with the big company, and a worldwide business resulted.

Marcellus Jacobs had a somewhat similar experience. As a Montana farm boy he took electrical courses in high school and later attended a special class in Kansas City. He was an avid reader of any books he could find on electricity, and while still in school he built and sold small radios operating on storage batteries.

The Jacobs ranch was forty miles from the nearest town and had to get along with kerosene lamps until Marcellus improved the situation with a secondhand engine, a little generator, and some old car batteries. He was nineteen when he experimented with linking the generator to one of the water-pumping windmills on the farm. The regular fan did not prove satisfactory, though Marcellus played around with it for three years. Then he decided that three blades of Sitka spruce were better than a whole circle of them, having got that notion after flying airplanes himself. His next problem was to find a means to take the pressure of the wind off the blades in a real gale, so he invented a governor for the machine. It worked so well that he built about twenty-five electrical wind plants for neighboring ranchers, buying the generators and towers from manufacturers.

This was the beginning of a successful commercial enterprise for Marcellus and his brother. The company was later moved to Minneapolis, where Jacobs went on improving the components of his machine. By 1933 he had perfected a design that was little changed for the next twenty-five years. His wind plants ran all sorts of household equipment as well as radios. They were known all over the United States. Today they are often sought by people trying to adapt them to modern uses.

chapter **15**

Changing Times

Numerous manufacturers undertook the production of small two- or three-bladed windmills for rooftops in the 1940's. These were a great help for families living in regions far from power lines. Sweeping in circles four to six feet in diameter, mills charged batteries not only for radios but also for a limited number of electric lights and for some household appliances.

Wind-driven generators require no special care except an occasional oiling. For this reason they can be erected in uninhabited spots. Air lanes across the American continent eventually were marked out with beacons consisting of tiny windmills that kept the lamps burning at night in high places. At daybreak a solar cell shut off the current and at nightfall released it again. This principle was also applied to lonely lights along the coastline, marking dangerous reefs and rocks.

Though the small-bladed windmills had solved some of the farmer's living problems, they did not help him operate larger equipment, such as refrigerators. He was still dependent to a certain extent on energy from oil, gas, or wood—until about 1950, when the Rural Electrification

Very few windmills of this vintage remain in the West. This one, pumping water into a tank in the tower, is at Edgewood, Washington.

Photo by Werner Lenggenhager

A few windmills of the conventional fan type are still seen. Early in the present century they were the commonest kind. This one pumps water for livestock at Mesa, Washington.

Photo by Werner Lenggenhager

Program brought the central power grid within reach of virtually every household in the United States.

This had a discouraging effect on the windmill manufacturers. Whereas in the late 1920's twenty-five companies had been marketing 99,000 small-type windmills annually, sales now dropped steadily until they touched bottom forty years later with less than 300. Out of more than 6 million that had been sold in the United States since 1850, only 150,000 were working in 1977, and they were mainly pumping water on ranches in Arizona, western Texas, and other arid parts of the Southwest. A few were being sold to generate electricity on a small scale.

Cheap fuel derived from oil had come into common use, and on the surface it appeared to spell the final decline of the windmill in any form. Yet unseen factors were at work altering the world's economy. Ecologists were talking about the planet's dwindling supply of natural resources. The coal and petroleum would not last forever. A new sentiment surfaced in the 1970's, especially among the young people —a desire to utilize energy sources that were free and inexhaustible. The sale of windmills picked up noticeably. They were purchased by nostalgic country dwellers, practical ranchers, and estate owners who missed the sight of whirling fans. Many cattle and sheep raisers had never forsaken them because windmills could bring water to widely separated herds, needed little maintenance, and ran by themselves. One man in Nebraska owned seventy windmills. And finally there were Amish farmers whose religion forbade them to employ electricity.

The vexing feature of wind energy had always been its unreliability and the difficulty in converting it into electricity and storing it in batteries, an expensive process. Electric generators have certain power ratings that must not be exceeded, so what energy they store lasts only a few days.

Engineers and scientists were challenged by the problems, feeling that an abundant natural resource was being wasted. Was it possible to harness the wind on a practical scale and make it work for an industrial society at a lower economic and environmental cost than fossil fuel or nuclear energy?

In widely separated places scientists considered trying out propeller-shaped windmills, not the little rooftop models but huge steel monsters weighing many tons, mounted on taller towers than had previously been used. It had been discovered that thin-bladed arms rotating at high speeds were most effective in generating electricity.

The machine that led the way became celebrated, for it was constructed long before the revival of wind power was being talked about. Back in 1934 Palmer C. Putnam built a house on Cape Cod and found the charges for electricity so costly that he began to think about obtaining power for his home from the wind, which blew steadily on the Cape. He investigated the subject and learned that the small-bladed units then being sold would not be sufficient for all the uses he had in mind. A neighbor told him about a barrel-shaped three-bladed Savonius rotor invented by a Finn, which the owner of a Long Island estate had erected. It seemed to be doing the work well, so Putnam inquired about other wind-driven generators that had been tried out in France, Russia, and elsewhere in Europe, but all of them were costly to install. Some, he learned, employed large, long blades, and he became interested in the results obtained. By this time he was thinking of a machine that would tie in its current with a regular utility system, so that at times when the wind died down there would be a backup source of electricity to feed the system. He succeeded in involving a number of professional men in his project, and in 1939 they formed a company to sponsor the erection of a tower about 150 feet high, with two huge eight-ton blades. A 1500-

kilowatt generator would be contained in a compact capsule-shaped room behind the wings, which could be reached by climbing a ladder on the four-sided iron-lattice tower.

First a suitable site had to be found. A peak known as Grandpa's Knob was bought from a Vermont farmer and a road built to it. The hill had an elevation of two thousand feet and was believed unlikely to have destructive ice storms. But it did have an abundance of good, steady wind.

Erection of the windmill was completed in August 1941. Tests proved it to be satisfactory, so that in October it was hooked into the lines of a Vermont utility company, the theory being that as long as the wind blew steadily, any excess power not needed by the windmill owners would go into the general supply and, on days of little wind, the utility company would take over. This was the first time anywhere wind power was fed into the high voltage line of a utility system.

Palmer's associates felt that if their giant machine was a success, they might go into the manufacture of a hundred units to be patterned after it.

The windmill ran without incident until February 1943, when a bearing in the mechanism failed. This was during the Second World War, when parts were scarce, and twenty-four months went by before the new bearing could be installed. The apparatus went back into power production on March 3, 1945, and for twenty-three days there was no trouble.

Then one night, when the wind was steady and the machine was operating smoothly, Harold Perry, the man on duty alone in the control room high aloft, was suddenly thrown face down on the floor and jammed against the wall. He knew something had gone wrong with the powerful blades, and realized that he must turn them off at once.

Perry struggled to his knees and was about to get up and head for the control panel when the vibrating monster threw him down again. There was no one to help him—if he was going to survive, Perry must reach that panel. He made a third try and this time succeeded in getting to the controls before another contortion of the mechanism floored him again. The horrifying movement stopped at once, and Perry was able to get down from the tower and contact John Wilbur, chief engineer on the project. Wilbur immediately phoned Putnam. "We've had an accident," he said, "but it could have been worse. We've lost a blade, but no one is hurt and the structure is still standing."

Wilbur wasn't too surprised that a blade had fallen off because there had been indications that it was cracking. The weakened steel had not been able to withstand the tension of centrifugal force amounting to several hundred thousand pounds. When it split, the great wing was thrown 750 feet and landed on its tip.

The night's experience had been a terrifying adventure for Perry, but it was very costly for the company. The break could not be repaired because of wartime conditions, so the windmill shut down for keeps. And although its sponsors were still in favor of this type of power and had learned how to correct any flaws, too much money would have to be spent for rehabilitation. They decided not to continue with the project, and the mill on Grandpa's Knob became a memory.

A lesson was learned by this small company exploring at the frontiers of knowledge, but though the men involved in the project succeeded technically, they could not afford to find out if they could succeed economically. The hundred units they had visualized never materialized. Today others are benefiting from their experience and making use of the knowledge they acquired.

chapter 16

The Danish Experiments

Denmark has always been a country that places great faith in wind power. It has some five hundred islands and is one of the windiest areas in Europe, with a strong breeze blowing nearly three hundred days of the year. By 1900, Denmark had about 100,000 windmills in such common use that they were called house mills. They threshed grain, milled flour, drained land, pumped water, sawed wood, and a few generated electricity. Before the turn of the century the government set up a royal commission to develop a wind-power plant especially adapted to the needs of the dairy industry.

The Danes still built windmills according to the old pattern, and although they put the new ones on high towers, they were topped with four sails.

When the First World War began, Denmark ran out of fuel, for it did not possess other sources of energy. Immediately a demand was created for more mills capable of generating electricity, and rural power stations were equipped with them.

After the war, Denmark, not wishing to be behind the

times, reverted to gasoline and diesel oil as soon as these commodities were again available. By 1920 only seventy-five windmills remained in operation. Danes were quick to point out that the wind did not work as steadily and was less reliable than fossil fuels for supplying electricity.

This attitude served the country very well in peacetime, but along came the Second World War, and now Denmark was in the same predicament as before. There was no recourse but to reactivate the windmills, and from 1940 to 1945, most of the nation's electricity was supplied by them. This time the country prepared to make wind power more efficient, and it started a research program to study this ancient resource, which had twice performed a much appreciated service in periods of stress.

Unfortunately, most people lost interest in the project when inexpensive hydroelectric power from Sweden became available. A few stalwarts retained their enthusiasm for wind energy and joined forces in an important experiment at Gedser on the island of Falster. First a study was made of wind velocity and direction, and thirty different blade designs were tried out, the purpose being to find the most effective way to capture power that could be transferred to an electric grid system. The Danes had advanced a step and were prepared to abandon fans and four wings covered with sails.

The investigation was extremely thorough, and it was not until 1957 that the results took shape in the form of a giant reinforced concrete tower topped with three huge blades and a cabin for the generators. Danes spoke of this impressive structure as the final result of eighty years of study to produce electricity by the wind. It was believed that, if this kind of windmill proved successful, electrical systems in other parts of the country could collect energy from a whole series of wind-powered plants.

The Gedser mill began operating steadily and met its builders' expectations in every way. But it couldn't compete with the low price of oil, and after eleven years, the mill was shut down in 1968. The great structure, more than eighty feet high, with its big propellerlike blades stood idle for ten years because the population preferred to have electricity made from imported fuels. It took the 1978 gas-and-oil troubles in the Middle East and the increase in prices of petroleum products to change the public's mind. The Gedser mill was started up again and, in the hope of stopping the drain on the national finances due to fuel costs, the government now plans to subsidize similar machines on other islands.

The whirling arms of such big windmills usually are connected to the generators either by gears, chains, or toothed belts. The Gedser transmission was by means of roller chains, and once the mill went out of service because a chain ran out of its sprocket.

The people in charge of operations were accustomed to seeing many inquisitive visitors from other lands. An American looking for ideas that could be applied in the United States gazed in astonishment at the automatic shut-off device for emergencies. Instead of some expensive piece of hardware, the Gedser mill employed the simplest kind of device. This is how the visitor described it: "A pipe comes up from the floor and bends around some equipment and on top of the pipe there is a cup in which sits a heavy, oversized ball. That ball is connected by a string to an old-fashioned switch on the wall. If the tower starts to vibrate, the ball rolls out of the cup and the string pulls the switch to stop the machine." Can you imagine anything less complicated?

Even larger than the Gedser mill is a newer one built by students and faculty of the Tvind Schools, a private learning center that receives some public funds. The establishment is comprised of a colony of 120 teachers and nearly 600 students who live on a 100-acre campus in West Jutland, ten miles from the seacoast. The educational program is unusual, for it includes the study of many national and worldwide problems. The students live in cooperative and communal groups and work together with the staff on various projects. One of these was a windmill for the school. At first they intended it to furnish heat only for the buildings, but after they learned more about the kind of machine under consideration they decided it ought to be able to provide light as well, which would make the colony self-sufficient insofar as energy was concerned.

Many outsiders became interested in the school's project and gave their time and knowledge freely to the experiment. Instructors contributed part of their salary to the purchase of parts, engineers donated their services, and almost all of the labor was volunteered. Aerodynamics experts working at the atomic energy agency furnished a design for the three blades, and a German professor suggested that they be made of fiberglass-armored polyester with plywood ribs. The group was unfamiliar with this material and experimented with it by building three fiberglass boats, which now supply fish for the school. The team then borrowed a portable military hangar from the Air Force for a workshop and set out to build the wind machine from as inexpensive materials as possible. The only thing the Tvindites had in common was that none of them had ever built a windmill before. Yet when they were finished they had the biggest one in the world at that time—170 feet tall.

It took nearly three years to construct it, and the tower was finally ready for operation in 1977. One of its remarkable features was that the equipment was completely controlled by computer. Observers were astonished that such an impressive accomplishment could have been conceived, organized, and mainly constructed by amateurs. Besides having all the power the schools needed, the Tvindites were able to sell electricity to the West Jutland network. The "Tvindmill" has served as a model for the rest of the world, showing what can be done with inexpensive techniques and local craftsmen, instead of hiring the costly services of high-technology engineering firms.

chapter 17

Wind Energy in the Modern World

If one were to prepare a calendar based on the erection of monster wind machines for producing electric power, the earliest entry would be Russia in 1931. Next would come Grandpa's Knob in 1941, followed by Costa Hill in the Orkney Islands in 1954, then one built in Wales in 1955 that was moved to Algiers, and the Gedser project in 1957.

There are many regions of strong winds on the Russian steppes, and the country has long used windmills, particularly for pumping water. Some of the earliest were oddly shaped and had a character of their own. Pictures show rows of post mills in Bessarabia (just across the border in Rumania) looking like farm sheds mounted on solid board bases and having six swinging arms.

As early as the fourteenth century Estonia and Latvia copied windmills from those then in use in western Europe, and rows of them became a traditional feature of the landscape both on the mainland and on islands in the Baltic Sea. Post mills in Estonia usually were set on stone foundations. Tower mills and smock mills embodied much fine handiwork and often were put together with more wooden pegs than nails. Though no longer in use, the remaining ones have become show places, and some are now in outdoor folk museums.

Russia's vast plains, deserts, and semideserts, although swept by ample winds, lacked supplies of electrical energy. Soviet scientists observed what other nations had accomplished with small windmills and generators and, following their example, produced some for charging storage batteries. These are used in such diverse places as field camps, shepherds' huts, rural clubs, tents, and farmhouses. The Russians erected a 100-foot tower at Yalta, near the Black Sea, which was unusual for its large blades made of roofing material. The blades drove wooden gears, producing wind energy that supplemented the electricity from a steam-powered plant at Sevastopol, twenty miles away. A daring achievement, it was another instance where makeshift components produced desired results. Damage during the Second World War put the windmill out of commission, however.

In 1978 the Soviet government announced plans for the erection of 150,000 wind-power facilities, among them approximately 4,500 towers to be completed in the Arctic regions by the end of 1980.

The British Isles are among the windiest places on earth, but it is difficult to find suitable sites there for the large, costly, and somewhat ugly structures necessary to capture enough energy to generate electricity. A government-appointed committee in the early 1950's found three places they thought feasible to try out aerogenerators, and the big windmills were erected in Orkney and North Wales and on the Isle of Man. The first one was on Costa Hill at Orkney, far north of Scotland, an exceedingly windy location, but so remote that it was difficult to make an installation, and the structure proved unreliable.

The government committee next turned its attention in 1955 to a wind machine at St. Albans in North Wales. It was built according to a design by Jean Andreau and had hollow blades with open tips. When it was in position and ready to hook its power into the lines of the British Electrical Authority, a strong whistling sound was produced by the air thrown out at the tips. This noise so disturbed the local inhabitants that they refused permission to keep the tower there permanently.

What to do with the costly experimental wind machine was a big question until someone suggested donating it to the Algerian Electricity and Gas Corporation, which was looking for a means of increasing its service. The windmill took a long boat ride to that Mediterranean country and found a new home on a breezy—and isolated—hill, where it ran until 1961.

The British energy committee's third experiment was finally successful. It went up on the Isle of Man in 1959, and there were no complaints until it lost a blade. After that, the committee ceased its efforts because of the greater availability of oil and the growing belief that nuclear energy would provide cheap electricity in the future.

Continuing our census of the great experimental windmills of that period, we should include a German one at Stotten, near Stuttgart, where Professor U. Hutter first used glass-fiber-reinforced plastic for blades in 1957. After that came three French mills, two in 1958, and the last in 1963. Professor U. Hutter had been involved in the Allgaier design, a smaller conventional horizontal axis mill in the postwar years. He was also an adviser when the Sandusky unit was constructed in the United States. Then once more the possibilities of wind power were forgotten—until the 1970's, when a reawakening took place all over the globe. Energy commissions and power-company engineers were not contemplating reviving old-type mills but were ready to give many kinds of modern rotors a trial.

What was good for one country was not good for another. New designs evolved, but in at least one instance old ones were made to serve. In Ethiopia an American Presbyterian mission introduced windmills with Cretan-style cloth sails to help people along the Omo River raise food crops. A visitor in 1975 counted nineteen mills of various types in operation in the village and five more on mission land. Eleven others had been completed and were to be installed as required for pumping. These included several Savonius rotors, which in a pinch can be made out of oil drums. The towers were only twelve feet tall, and most of the sail windmills had eight arms made of black water pipe with Dacron sailcloth attached. It was lucky for the poor farmers that the mission obtained a whole roll of this imported fabric. No mill owner left his mill unattended or with the sails on at night. They had to be taken in and carried back to the village, otherwise they would be stolen. This was not an easy way to end a day's work, for most villagers lived an hour's walk from their cultivated plots.

An effort was made to interest India in windmills for irrigation, but the people were not receptive to the idea; they were accustomed to lifting water by means of bullocks or diesel pumps. It did not matter that the animals had to be fed and their working life was less than ten years. Nevertheless, some tests were carried out in the windiest parts of the country, and in 1973 a large sail-wing mill with a single blade was constructed near Madurai to furnish electric power.

A willingness to try out wind power again exists all over the globe today. Modern mills have been introduced in Colombia and Haiti. One was tried out at Mogador, Morocco. Curaçao has three thousand American windmills in use for pumping water. Israel has tried out windmills at an isolated place on the Red Sea coast. Switzerland and Australia both manufacture windmills. There is an experimental large-bladed mill in use in Hungary, and there are wind turbines in Japan. The latter country employs some at lighthouses on isolated islands. In 1975, France was producing the largest commercially manufactured electrical unit.

West Germany is also building several types of windmills suitable for central Europe. A large one erected on the island of Sylt in the North Sea successfully converted 60 percent of the wind's force into electric power. That is quite a good record. Encouraged by this experience, Germany will erect a still larger mill north of Hamburg with a 300-foot-high tower.

The latest one planned will be in the state of Schleswig-Holstein and is called Growian (for Grosse Windenergie Anlage—Big Wind Energy Facility). The two-blade Growian rotor is 300 feet in diameter and mounted on 450-foot columns. It will stand as tall as the Cologne Cathedral, and the Germans claim it will be the world's largest. It will be the forerunner of four thousand rotors grouped in series along the North German coast.

New Zealand, Tanzania, Japan, and Sweden are all studying wind power. One finds it in use in the most unexpected places. For instance, in New Guinea wind generators were installed for telecommunications by the Department of Posts and Telegraphs. Thailand and Taiwan both manufacture small mills with wooden blades. Sri Lanka (Ceylon), Antigua in the West Indies, and Kenya in Africa have all given windmills a trial. They have been used as navigation aids at the offshore oil-and-gas production platforms in the Gulf of Mexico. Long before that they were associated with lighthouses in such widely separated lands as Italy, Canada, Yugoslavia, Australia, and France. One small Australian machine has powered a lighthouse in Sarawak, Borneo, for twenty-five years. Some have been installed at remote weather stations, such as the vertical mill Canada erected on the permanent ice in the Beaufort Sea.

Another project under study is a wind-driven tidal mill that extracts hydrogen and other chemicals from seawater. Storage of energy can be achieved through hydrogen gas obtained from wind-generated electricity. The National Aeronautic and Space Administration a few years ago proposed to produce hydrogen gas from river water, compress the gas, and store it for use during windless periods. Hydrogen gas can also be used in the manufacture of fertilizers.

The wind-driven tidal mill has been proposed for the Bahamas and the Virgin Islands, a region accustomed to seeing quite different wind-driven machines. There tourists can visit various old-fashioned tower windmills, which used to grind sugarcane on plantations. One has been completely restored at Whim Greathouse on St. Croix, and the blades of the mill still turn.

In the Bahamas, where the trade winds blow constantly, it was suggested that delta wings be erected on the roofs of large buildings to collect wind energy for industrial uses. Wind increases at high points above the ground, and this has prompted one American expert to suggest mounting windmills on the roofs of skyscrapers.

Two Australian engineers have come forth with an extraordinary idea. They want to send up large unmanned gliders carrying wind turbines and tether them when they reach an altitude of six and a half miles. At that height the jet stream, a band of fast-moving air currents swirling around the earth, can supply three thousand times as much energy as ground-level winds. The electricity produced by the whirling wings would be transmitted to earth through the cables that anchored the gliders. The glider would be designed to return to the ground if the wind dropped too much. Should the cable break accidentally, radio controls would bring the craft back to earth, while its fall would be slowed by parachutes.

The engineers are quite serious about this plan and have done much experimental work with gliders, which, they claim, are especially suited to Australia, where high-altitude winds abound over about a third of the continent.

chapter **18**

Great Mills in
the United States

The United States did not take a serious interest in wind energy on a national scale until 1972, when it launched an investigation to assess its possibilities. It was shown that by the year 2000, 18 percent of the nation's electricity requirements could be furnished by wind power, provided that the means with which to accomplish this were available.

Other countries already had awakened to the need for power from something other than oil. The Grandpa's Knob project had proved that wind could provide energy but that windmill was no longer operating. It—like the British experiments—was a victim of the belief that atomic energy would ultimately solve our forthcoming power problems. But now we are beginning to realize that there are many problems connected with atomic power.

The next effort to generate wind energy on a large scale in the United States was made by Percy H. Thomas in Vermont, who designed a tower with two turbines at the top and tried to interest Congress in the project. The Korean War occupied that body's full attention at the time, and so the proposal was set aside. Thomas put a great deal of effort

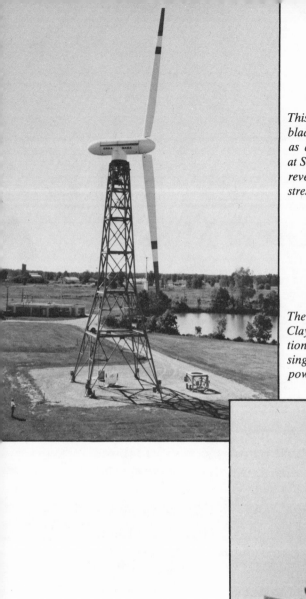

This was the first large-scale, two-bladed wind turbine built by NASA as an experimental prototype. Operating at Sandusky, Ohio, since 1975, it has revealed much information about stresses and other problems.

Courtesy NASA

The old in contrast with the new at Clayton, New Mexico. The conventional windmill was designed to serve a single farm, and the big turbine generates power for community use.

Courtesy NASA

into the plan, but he never had the satisfaction of seeing it carried out; the tower was never built.

Another engineer, Professor W. Heronemus of the University of Massachusetts, a former Navy captain and designer of nuclear submarines, suggested using ocean-based wind turbines off the coast of Maine. The state of Montana offered some plans along the lines of the Gedser mill, and studies made at Oregon State University tried to apply the Grandpa's Knob design to the Oregon coast.

Finally, late in 1974, some action was taken. The National Aeronautic and Space Administration and the National Science Foundation together awarded contracts for very large wind-energy systems that would generate electricity. The windmills would have wings two hundred feet in diameter, and each would be able to produce electricity for 100–200 private houses.

The first experimental model was erected at Plum Brook, near Sandusky, Ohio. It is operated automatically by a computer that monitors the wind and orders the turbine to move in the right direction to orient itself. The computer can also shut down the entire mechanism if the wind is too strong or some part is malfunctioning.

At first, in spite of being equipped with the latest control mechanisms, the windmill seemed to have a serious flaw, and the blades were almost destroyed by excessive vibrations during the early months of operation. Finally the designers identified the mysterious cause of the trouble. The tower had been furnished with a more elaborate stairway than was customary on such structures, and the steps created what engineers described as a "wind shadow." When a simple ladder was substituted, the windmill was restored to good health.

Four years later another giant of the same design but double the capacity was erected at Clayton, New Mexico. It

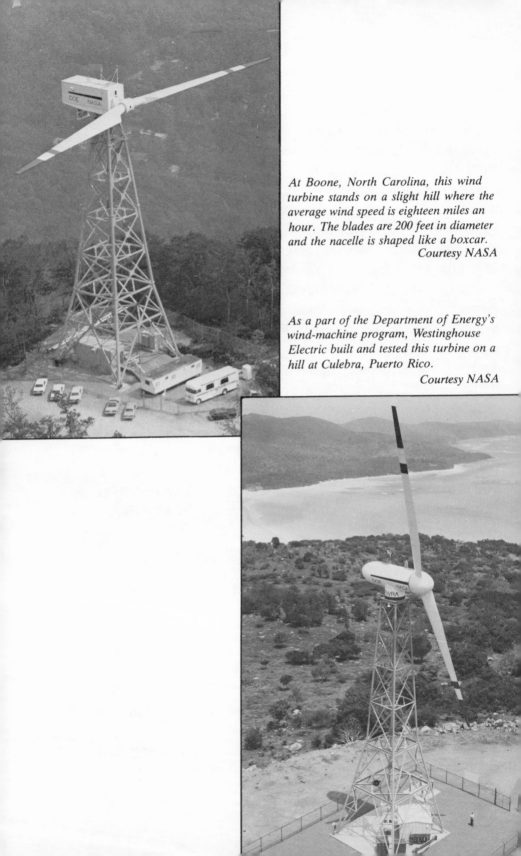

At Boone, North Carolina, this wind turbine stands on a slight hill where the average wind speed is eighteen miles an hour. The blades are 200 feet in diameter and the nacelle is shaped like a boxcar.
Courtesy NASA

As a part of the Department of Energy's wind-machine program, Westinghouse Electric built and tested this turbine on a hill at Culebra, Puerto Rico.
Courtesy NASA

has 60-foot-long steel blades atop a 100-foot steel tower and is expected to furnish power whenever the wind exceeds twelve miles an hour (which it does most of the time at that place).

A third windmill rose in 1979 on Howard Knob, a thousand feet above the town of Boone, North Carolina. Each of its blades is a hundred feet long and weighs nine tons. Like the Grandpa's Knob mill, it has a nacelle, an enclosed room for generators and instruments, 140 feet above the ground. It is computer-controlled, and whenever the wind blows less than eleven miles per hour or more than thirty-five, the blades are feathered (turned to offer minimum drag), and the system shuts down.

Official dedication of this monster was a great occasion for the residents of Boone, and they turned out bearing paper pinwheels and hundreds of small balloons. When the wind blows twenty-five miles an hour, this mill is supposed to satisfy the electrical needs of 500 one-family houses. It is connected into the General Electric Company system. Regarded as a model of things to come, it is hoped that by the year 2000 there will be thirty thousand such windmills, which could supply at least 2 percent of the nation's needs.

Large aerospace and electrical-equipment companies are doing most of the testing and preparation on the federally sponsored windmills. Lockheed manufactured the blades at Clayton, and Boeing supplied those at Boone. Other windmill installations have been made on Culebra Island, Puerto Rico, and on Block Island in Rhode Island. The Navy has a windmill built by the Grumman Company at the Marine Corps station at Kaneohe, Hawaii.

Another large mill stands on Cuttyhunk Island off the coast of Massachusetts. Privately owned, it is sponsored by Allen Spaulding, Jr., who became interested in wind power in 1974, when he installed a water-pumping mill on his property. After studying many designs he concluded the Gedser mill offered the soundest approach. He organized a company with several other men who had engineering skills, and together they directed its construction. The town of Gosnold on Cuttyhunk Island benefits from the surplus electricity the mill generates.

Not all the new American mills are on such an ambitious scale. Take, for instance, the one at a school in Waitsfield, Vermont. The school's director thought that since the world is facing an energy crisis, it would be good for the children to learn how the wind could help, and so they proceeded to build a windmill. When the structure went up they made models of it. They interviewed people about it, kept records of the mill's performance, and checked the machine daily. They found out how all the parts worked and learned some wind-power history.

An architect living at Point Richmond, California, in 1974 had his firm search for a windmill suitable for his new home overlooking San Francisco Bay. After nine months of study his associates recommended a type of mill serving ski resorts in remote areas of the Swiss Alps. The turbine was expected to provide all the power needed for several houses, which the firm was interested in building. Its storage batteries were intended to provide enough energy for three windless days. Some residents were afraid of what the windmill would do to their view, but when it was erected it turned out to be much more attractive than the unsightly power lines.

A rather unusual place for a giant turbine to go up was at the Dorney Amusement Park near Allentown, Pennsylvania. It was erected by an independent inventor, Terry Mehrkam, who coupled pieces of industrial machinery together and placed four large blades on the tubular tower, which has an access ladder. Surprisingly, when he finished,

This wind turbine at Block Island, Rhode Island, is another furnishing power for a community. Its blades have a diameter of 125 feet. The nacelle is streamlined.
Courtesy NASA

the mill produced more electricity than the one at Clayton built by experts.

Not such a great departure from the old concept is the twelve-foot wind turbine, a round wheel with many spokes, which sits on a 13,000-foot peak near Copper Mountain, Colorado. In 1977 it was the first of its kind in the United States supplying power for a transmitter that gathered and rebroadcast television signals to the community. Two years earlier Oak Creek, another mountain town in the same state, employed a windmill to operate an FM broadcasting station. A few times it had to go off the air for lack of wind.

There is not always easy acceptance of windmills by those who live nearby. A resident of Mechanicsburg, Pennsylvania, wanted to erect one in his backyard and use it to supply heat and light for his house. He had to give up the plan because the neighbors complained.

Usually the objection to the tall turbine towers is that they can interfere with television reception. Efforts to install the turbine on Block Island met with continued hard luck. The New England Telephone Company made the first attempt in 1976, but the blades blew off in a stiff breeze. Another effort in 1979 seemed ensured of success until the residents complained that the whirling wings interfered with television reception. As a consequence the operation had to shut down while the Department of Energy, which was sponsoring it, spent $700,000 to install a cable-television system on the island.

chapter 19

Giants in
the Northwest

Windmills have changed from the quaint to the ultra-modern in the Pacific Northwest. Oregon and Washington had their own version of mills a century ago, when farmers constructed wooden towers three or more stories high and topped them with one of the popular multibladed fan wheels of that day. Some of the towers were shingled, all were painted, and usually there was gingerbread trim. Always there were windows, and frequently the top was finished with a rail like the kind found on an observation platform. They were usually built over a well and had a durable large redwood tank on the top floor.

These mills went out of style several generations ago, and the few that remain standing no longer have a fan on the roof; instead they house an electric pump.

The region abounded in hydroelectric power, and there was no thought of ever going back to the windmill until the threat of limited fuel-oil supplies inspired scientists at Oregon State University to place wind-measuring instruments on bluffs along the coast and on a television tower up the Columbia River. Their research into wind patterns,

An artist's conception of the monstrous Boeing wind machines of the type slated for the Goodnoe hills near Goldendale, Washington. They are the largest yet designed and have a 300-foot blade diameter.　　　*Courtesy NASA*

which began in 1971, turned up interesting facts, such as that the highest sustained wind on record in the region was 113 miles per hour at the mouth of the Columbia in January 1921.

That was not an ideal location for building windmills, because they must be placed where there are steady winds, not high-velocity gusts. Farther up the Columbia Valley are locales that do have steady winds, and the National Aeronautics and Space Administration decided in 1979 that three large wind turbines manufactured by the Boeing Company should be constructed in the hills beyond Golden-dale, Washington. These windmills are the largest ever made: the blades are 300 feet in diameter and are mounted on 200-foot steel towers equipped with an inside lift. Each generator is expected to produce 2,500 kilowatts of electricity, depending on the wind, and supply enough power for 750 homes. The Boeing Company is working on the assumption that large windmills have become economical for the nation's electric utility companies. The choice of a location for the first three was influenced by the existing hydroelectric power source, which could be depended upon when the winds dropped in force. The Bonneyville Power Administration will be in charge of the operation and will integrate the output of electricity into the Klickitat County grid. The area has the advantage of being close to dams that can store water when windmills are producing energy and later release the water at times when another source of power is needed.

Another giant tower, 262 feet above the prairie, is planned for south central Wyoming, to be in operation in September, 1981. It is being considered as possibly the first of fifty large wind machines at this remote region in the interior and may become the nucleus of an electricity-producing wind farm.

chapter 20

Looking into the Future

By 1975, forty federal projects were underway in the United States, sponsored by the government in its effort to employ the wind as an alternative source of energy. The study is concentrating on two types of windmill: those with horizontal-axis, two-bladed propellers, and, to a somewhat lesser extent, on those with high-speed vertical-axis units. Hope has been expressed that once the government project determines the most efficient way to use windmills, private enterprise will be interested enough to carry on.

Researchers have already taken a census of favorable sites for mills and of the periods when they can serve most efficiently. They have made wind maps of the world, showing the best and poorest regions regarding the amount of wind available. The best places in the United States are mountaintops, ridges, the western Great Plains, and off-shore sites at sea. The East and West coasts, the areas around the Great Lakes, and the Cascade, Rocky, and Appalachian mountains offer many promising sites. The problem is that the windiest places are often also the most scenic. Usually they are remote from population centers, and some are in areas where access roads or bridges would have to be built in order to reach the mills.

Other factors that have to be considered are soil conditions for supporting the tower, the effect on migratory birds, the prevalence of gusting and cyclonic storms. Before starting work on the new windmills on the Columbia River, an engineering crew flew kites at the site and examined the topography, vegetation, and wildlife.

Wind at any given spot is often more powerful than shown by conventional weather recorders, and the towers have to be able to withstand the highest expected gust. One expensive imported windmill was blown down in a mountain storm. Ice, dust, lightning, metal fatigue, and stray bullets from hunters or boys wanting a target present other problems.

Searching for locations where wind energy is abundant is something like prospecting for minerals. It has been said the nation needs to encourage prospecting for such sites as it once did for uranium deposits. Winds are not always the same every year. Velocity is sometimes increased if wind has to flow through notches or around obstacles. A survey of wind-energy sites in Hawaii revealed that the best were in slots in the mountains or around the corners of islands.

Windmills are practical only where the wind blows seventeen miles or more per hour at least 80 percent of the time. Any windmill has three critical speeds, the cut-in speed below which it is not worthwhile to run it, the rated speed for full output, and the furling speed, at which it must be shut down to avoid damage.

Although the average person has yet to become accustomed to the look of today's mammoth windmills, there are other shapes not so forbidding, but rather startling. Scarcely recognizable as being of the same family is the Darrieus rotor, which has appeared in recent years and comes in all sizes. Many small ones can be seen in Denmark on rooftops and water towers. Patented by G. J. M. Darrieus in 1927 in France, this type of mill was forgotten in America until two Canadians, Raj Rangi and Peter South, rediscovered the principle during the late sixties. The machine stands upright and resembles an eggbeater with two curved blades. They turn no matter which way the wind blows. Some sold in 1974 for use on housetops had 14-foot-high blades curved like hunting bows. These were attached at top and bottom to a 15-foot shaft.

Lately, much larger ones have been envisioned, and Canada has erected a huge Darrieus on the Magdalen Islands in the Gulf of St. Lawrence.

On a July morning in 1978 a crew arriving at work when it was still in the experimental stage found the turbine was already turning. (The eggbeaters have to be started.) The main brake had been disconnected and the aerodynamic brakes (flaps hinged on the blades) were banging open and shut. The turbine picked up speed and was spinning at twice the maximum rate for which it was designed. A blade struck one of the supporting cables and down came the entire rotor, then the largest and strongest one that had ever been built. This didn't discourage the Canadian company, Hydro-Quebec, and it plans another, 108 meters high, to be in operation in 1983.

The United States Department of Energy at its Sandia Laboratories in Albuquerque tried out an eggbeater with three blades eighty feet long. Now the Aluminum Company of America has constructed a vertical-axis wind turbine, as they are officially known, that stands 123 feet tall and has three hollow aluminum blades, each 29 inches wide, turning in a circle 82 feet in diameter. Erected in the summer of 1980 on the Oregon coast, it delivers 500 kilowatts of electricity in winds 12–35 miles per hour.

This huge Darrieus, or eggbeater-type, wind turbine has been erected at Sandia Laboratories at Albuquerque, New Mexico. Another at the same experimental station has three blades. *Courtesy Sandia Laboratories*

Sweden and Denmark have considered erecting groups of wind machines like these, but people there have questioned how they will look on the landscape. Unlike the early windmills, there is nothing picturesque about them.

Americans, in making plans to generate electricity on a large scale, think in terms of windmill farms. Vast amounts of unused area, either land or ocean, would be required for mills of commercial capacity. Spacing must be considered, and at least one acre is needed for each unit. Slender concrete towers seem to be the most acceptable design for supporting the blades.

Two or three mills arranged along a mountain ridge might not make much difference to the landscape, but several hundred in a section of the Kansas wheat belt would be quite another matter, even though they produced the equivalent of a nuclear-power plant. In California a private firm in 1979 began to build twenty windmills in Pacheco Pass, eight miles south of San Francisco, to supply power for a thousand people. If this proves a successful enterprise, several hundred mills will be erected on a state-owned site.

In 1972, Professor Heronemus proposed that windmills —thousands of them—be floated off the Atlantic coast on offshore platforms to help alleviate the power shortage. The windmills would be towed to their anchorage and towed back to port for repairs when necessary. Each group would have one transmission line to shore. Britain already has explored the idea of placing clusters of about two hundred windmills in areas of shallow water.

Heronemus also suggested using windmill energy at sea combined with the electrolysis of water to create hydrogen for use as fuel. He visualized "farms" of windmills along the Aleutian Island chain with tankers to carry the liquefied gas to California.

In 1974 the professor had another vision—of a network of 300,000 giant windmills, each 850 feet tall, spaced one per square mile across the Great Plains from Texas to the Canadian border. He saw them with propellers 500 feet in diameter and believed they could utilize 59 percent of the wind power in the area. Such a system, he felt, could produce one trillion kilowatt hours of electricity per year, more than half the nations's needs.

Another American engineer talks of monstrous tornado towers into which the wind would spiral by means of louvers. But tornadoes are noisy, and such machines would need acoustic refinement.

Rotor turbines have taken other unusual shapes. Captain Sigurd Savonius, a Finn, in 1924 invented the Savonius rotor, which looks like a barrel hoisted on a pole. As mentioned before, wind machines have been made by simply cutting oil drums in half and attaching them in the shape of an *S* to a central post.

Then there is the giromill with its open-square vertical blades that rotate like a merry-go-round. A large one, ninety feet tall, was erected at Rocky Flats, Colorado. The three 30-foot blades were mounted on top of a steel lattice tower. The Department of Energy has been trying it out as part of the rural and farm-use program.

The cycloturbine is another variation of windmill. An Israeli researcher encased a wind turbine in a shroud somewhat like a circular airplane wing which, he claims, has tripled the energy of other mills. Strangest of all is the Madaras system, which consists of a series of cylindrical rotors, each mounted on a wheeled cart that is pushed around a circular track by the wind. Generators are attached to the wheels of the carts.

There are spiral windmills, squirrel-cage windmills, and hinged-vane windmills copied from a drawing nearly 400 years old. There are many brands of blade windmills, each somewhat differently shaped. Princeton University has pioneered the use of one called a sailwing.

New kinds of blades have to be designed for different conditions. To be most efficient, they must have both twist and taper. They are the most expensive part of the big new machines. On the tower at Clayton, cold and snow and an ice buildup have to be reckoned with, whereas in Puerto Rico the salt and humidity effects have to be considered. In the government's experiments with sophisticated designs, some mills have run only a week and then broken down. Others have been operating smoothly, and engineers hope that the new large turbines will last for thirty years. They theorize that if the mills were spaced out over a large enough area, such as Heronemus suggests, there would always be some wind in the lineup, even though some of the wings would be idle at times.

Experiments up to now have been extremely costly, but they have a common aim—to design wind machines with long life that can be manufactured economically and marketed by the power industry.

Invention follows need. Imaginative uses for wind energy are being developed while the experimenting goes on. It has been applied to desalination of saltwater to supply moisture for dry lands. In places much brackish water lies beneath the surface of the earth and could be directed toward industrial, municipal, or agricultural purposes if the salt were removed. With a windmill driving a high-pressure pump furnishing electric current, the brine can be forced through a plastic membrane that separates the salt from the potable water.

An East German farmer is reported to have found an extraordinary use for wind power. He drives his tractor with it, having mounted rotors on the machine. They take the place of gasoline.

Today the most basic and elementary questions are being asked about wind power, and men are studying the ancient designs in planning new machines. Thus we have windmills with upright blades rotating around a central shaft set in the ground, much as the Iranians built them many centuries ago. The modern version has an electrical generator conveniently located at ground level. Windmills of this type have been found to be the least costly to build.

New manufacturing companies have come into existence, and new publications on wind energy have appeared, proving that the public is interested in the subject. Designers everywhere are working on new styles. There are also many people in the United States who build windmills for their own personal use as a hobby, often out of old parts. Some of these windmill enthusiasts tour the countryside collecting parts of old mills, even if it means climbing a 45-foot tower to bring the mill down. If they are lucky they may find one stored in a barn.

Wind power is an environmentally clean resource. It produces no waste, especially no radioactive waste with its storage problems. It does not cause cancer or black lung, nor does it mar the landscape like strip mining does. It does not pollute the atmosphere and create smog like the internal-combustion engine. And whereas the latter operates on an irreplaceable fuel, wind power is inexhaustible.

But the wind must be used where it occurs. Unlike water, the air stream cannot be channeled or held back to store energy. It constantly changes direction and speed. Today the most economical way to use wind power is to combine it with an electric grid as a backup source of energy. This saves other types of fuel when the wind blows. Eventually some improved kind of battery may make it possible to store surplus wind energy for future use. Ways to reduce the costs of installation of the big mills must also be found.

As the standard of living improves, the population consumes more energy. If the wind should provide only 1 percent of the nation's energy needs, it could reduce the cost of importing petroleum by millions of dollars. The aim is to do better than that and go as far as producing 5 percent of our power, which would represent a staggering saving of precious fossil fuel.

The problem today is how to increase the windmills' performance and decrease the cost of building them. It is a matter of economics and of finding suitable sites. There is also the question of how to make the public accept the idea.

Few people take wind energy seriously; they do not think it will ever make an important contribution to the nation's needs. And often there is outright rejection of the idea.

The Netherlands is an example. Its historical windmills are of no use in terms of modern technology, but when it was suggested that the kind be installed from which much of that country's electricity could be derived, a government commission decided that there was no region where the siting of such towers would be acceptable. They just wouldn't fit into the environment. The gaunt, massive structures are not picturesque like the old mills, and the Dutch were afraid they might hurt their thriving tourist industry.

Resentment to the towers' intrusion on the landscape in other parts of the world has also been expressed. Even some of the Danes feel that way about it, although they have probably seen more of the modern mills than any other people. It isn't only the rejection of the visual appearance of the giant mills; some people fear that the great rotating blades might cause serious injuries to the population if they were damaged and fell.

In spite of this opposition, a return to a certain amount of wind power may be inevitable. Windmills are the least damaging to the environment of any energy source. There is a growing awareness in many lands that ultimately all essentials of life must flow from replenishable or recyclable sources. As Stewart L. Udall, former Secretary of the Interior, summed it up, "Our inventors will have to build us machines that use, not abuse, the unearned gifts of nature."

Further Reading

BOOKS

Ball, R. S., *Natural Sources of Power*. New York: Van Nostrand, 1908.

Batten, M. I., *English Windmills*. London: Architectural Press, 1930–32.

Bennett, Richard, and Elton, John, *History of Corn Milling*. New York: B. Franklin, 1898–1904.

Brangwyn, F., and Preston, H., *Windmills*. New York: Dodd, Mead, 1923.

De Camp, L. Sprague, *The Ancient Engineers*. Garden City, N.Y.: Doubleday, 1963.

Dennis, Landt, *Catch the Wind. A Book of Windmills and Windpower*. New York: Four Winds Press, 1976.

Mother Earth News Staff, *Handbook of Home-Made Power*. New York: Bantam Books, 1974.

Muir, Sir William, and Weir, T. H., *The Caliphate*. Edinburgh, Scotland: J. Grant, 1924.

National Institute of Agricultural Engineering, *Windmills for Generation of Electricity*. Oxford, England: Oxford University, 1933.

National Science Foundation, *Wind Energy Conversion Systems —Second Workshop*. Frank R. Eldridge, ed., 1975.

Nicholson, John, *The Operative Mechanic and British Machinist*. London: Knight and Lacey, 1825.

Putnam, P. C., *Power from the Wind*. New York: Van Nostrand, 1948.

Reynolds, John, *Windmills and Water Mills*. New York: Praeger, 1970.

Sarton, George, *Introduction to the History of Science*. Baltimore:

Williams and Wilkins Co., 1927–1948. (Published for Carnegie Institution.)

Simmons, Daniel M., *Wind Power.* Park Ridge, N.J.: Noyes Data Corp., 1975.

Stokhuyzen, Frederick, *The Dutch Windmill.* Bussun, Holland: C. A. J. van Dishoeck (n.d.).

Storck, John, and Teague, Walter Dorwin, *Flour for Men's Bread.* Minneapolis: University of Minnesota Press, 1952.

Wolff, Alfred R., *The Windmill as a Prime Mover.* New York: Transactions of the American Society of Mechanical Engineers. J. Wiley & Sons, 1885.

Wulff, Hans, *Traditional Crafts of Persia.* Cambridge, Mass.: MIT Press, 1966.

ARTICLES

"Aussie Wind Power Is for the Birds." *New Scientist* (March, 1979).

Baker, Robert W., "Windpower Potential of the Northwest Region." *Power Engineering* (June, 1979).

Black, Ted, "Putting the Wind to Work." *Design Engineering* (January, 1980).

Cadwallader, Edgar A., and Westberg, John E., "Wind-powered Processing." *Chemtech* (April and May issues, 1979).

"Clayton Turns to Wind Power." *New Scientist* (February 16, 1978).

Fuller, R. Buckminster, "Wind: the Answer?" *Solar Engineering* (November, 1979).

Gipe, Paul, "Mehrkam's Windmills." *Popular Science* (April, 1979).

Hamilton, E. P., "Some Windmills of Cape Cod." Newcomer Society for Study of History of Engineering and Technology, *Transactions,* Vol. 15 (1934–35).

Johnson, Lee, "Wise Wind." *Rain* (April, 1977).

Kilar, L. A., "Wind Blows Anew." *Power* (May, 1979).

Lapin, Ellis E., "Economic Competitiveness of Windmills." *Energy Conversion,* Vol. 16 (1977).

Melaragno, Michele, "Windpower." *Engineering* (January, 1977).

Merriam, Marshall F., "The Tvindmill." *Rain* (January, 1978).

———, "Wind Energy for Human Needs." *Technology Review* (January, 1977).

Olgaard, P. L., "Technical Note." *Solar Energy,* Vol. 22. Pergamom Press. (Printed in Great Britain.)

Power, H. M., "Windmills on the Mind." *Electronics and Power* (April, 1979).

Reddoch, Thomas W., and Klein, John W., "No Ill Winds for New Mexico Utility." *IEEE Spectrum* (March, 1979).

Smith, R. Jeffrey, "Wind Power Excites Utility Interest." *Science,* Vol. 207 (February 15, 1980).

Soderholm, Lee H., "Rural Use of Wind Power to Conserve Energy Resources." *IEEE Transactions on Industry* (November–December, 1978).

Strasser, Sylvia, "Harnessing the Wind." *World Press Review* (April, 1980).

Symonds, Mary, "Back-up Energy Blowing in the Wind." *Offshore Engineer* (February, 1979).

Taylor, R. H., "Wind-power Research and Development in the United Kingdom." *Electronics and Power* (July, 1979).

Tewari, Sharat K., "Economics of Wind Energy Use for Irrigation in India." *Science,* Vol. 202 (November, 1977).

Torrey, Volta, "Blowing Up More Kilowatts from Wind." *Technology Review* (February, 1980).

"United Kingdom Supports Vertical Axis Windmills." *New Scientist* (January 12, 1978).

Wade, Nicholas, "Windmills: the Resurrection of an Ancient Energy Technology." *Science* (June, 1974).

Wailes, Rex, "Windmills of Eastern Long Island." Newcomer Society for Study of History of Engineering and Technology, *Transactions,* Vol. 15 (1934–35).

Warne, D. F., and Calnan, P. C., "Generation of Electricity from the Wind." *IEEE Review* (November, 1977).

"Wind Energy, Net Energy and Jobs." *Rain* (May, 1978).

"Wind Energy Update." *Rain* (July, 1977).

Wind Power Digest (Winter 1975 through Winter 1977).

"Windmills." *Oklahoma Today* (Autumn, 1978).

Yen, James T., "Harnessing the Wind." *IEEE Spectrum* (March, 1978).

Index